THE BOREAL DRAGON

OTHER BOOKS BY KATHERINE BITNEY

Poetry:
Firewalk
While You Were Out
Heart and Stone
Singing Bone

THE BOREAL DRAGON

Encounters with a northern land

Katherine Bitney

WOLSAK
& WYNN

© Katherine Bitney, 2013

No part of this publication may be reproduced, stored in a retrieval system or transmitted, in any form or by any means, without the prior written consent of the publisher or a license from The Canadian Copyright Licensing Agency (Access Copyright). For an Access Copyright license, visit www.accesscopyright.ca or call toll free to 1-800-893-5777.

Cover design: Julie McNeill
Author's photograph: Mandy Malazdrewich
Typeset in Stempel Garamond
Printed by Ball Media, Brantford, Canada

The publisher gratefully acknowledges the support of the Canada Council for the Arts, the Ontario Arts Council and the Canada Book Fund.

Wolsak and Wynn Publishers Ltd.
280 James Street North
Hamilton, ON
Canada L8R 2L3

Library and Archives Canada Cataloguing in Publication

Bitney, Katherine
    The boreal dragon : encounters with a northern land / Katherine Bitney.

Essays and poems.
ISBN 978-1-894987-69-1

    I. Title.

PS8553.I8776B67 2013   C814'.54   C2013-901095-5

TABLE OF CONTENTS

Spring
March	11
April	14
May	16
The Green Dragon: Writing the Boreal (Section One)	18
Does Nature Have Rights? Ethical Implications in Ecology	28

Summer
June	51
July	54
August	57
The Green Dragon: Writing the Boreal (Section Two)	60
Eating Other Bodies: Some Ethical Considerations	74

Fall
September	85
October	88
November	91
The Green Dragon: Writing the Boreal (Section Three)	93
Retelling the Story of Nature: How to Restore an Ethos	109

Winter
December	125
January	126
February	128
The Water Project	129
Questions Toward an Ethics of Violence	143

Second Spring
Forest Diary, Falcon Lake, Manitoba	159

Acknowledgements	167
Author's bio	168

*This book is dedicated to my beloved granddaughters,
Julia, Jasmine and Jessica Bachand,
who hold the future in their hands.*

Spring

MARCH

Great white lion of March pads in, snow and snow. We are out with shovels again. Again. Last minute of winter, a little water for the earth, the plants, the creatures, come spring. Body shifts again to meet it.

Light snow, dull sky, a dusting of frost on the spruce trees. Spirits everywhere. And music.

Plus five degrees Celsius, snow melting, little rivers running under the soft snowbanks, into the gutters and drains. We have spring, early, at least for a while. Listening for the geese now, flying north, making their usual stop at the river.

Vee of geese honking to the west, a young straggler calling wait! wait! Crows squabbling on the boulevard, from the bare trees, whose lover is whose. Snow half-eaten by the sun. How dare it be so early, spring?

Do equinoxes mean anything here? Tipping of light for the birds going north or south. They move along the rivers, the moon. Currents in the land.

Early morning, high wind and warm air, and we are at the door of spring. Tomorrow the whole earth stands in balance, night and day, light and dark. New Year for Persians, Afghans. Turn of the great wheel of light.

A chickadee drops out of the sky.

Blessings of the vernal equinox! Here in the northern hemisphere we turn, ease into the long light, the big days of spring. Geese, songbirds wing back from their wintering in the south to feed and breed in the northern forests, the prairie, the tundra, singing. Robins hunt city gardens for food, nesting sites. Whole planet in unity, in perfect balance of day and night, dark and light. What will we do with this understanding?

Now there is the garden bunny, colour of earth, colour of last year's leaves. Colour of old wood. Colour of dry grass, of tree bark. Of wintered cabbage stalks. Of paving stones. Hunting for a place to hide a baby, come the new green leaves.

There is no land more sacred than the land you stand on.

Plus ten degrees Celsius, sun bright. A day lily already sending up little green shoots next to the chimney. Got yelled at by crow after crow as I passed under the trees. This is spring, fast and furious.

Shoulda been working indoors, but oh, could not. Went out to play, feet in wellies, spade in hand to crunch the rotting ice. Crows patrol the skies importantly in sixes and sevens; merlin on the wing, squealing. Chickadees and sparrows squabbling in the trees. Sun, oh sun, burning ice to water. Smoke of dragon's blood rising for the spirits of the land.

Cold, rainy this morning. Plants will rejoice, hold back their budding and pull in the water. Honeysuckle already leafing; hawthorn, cherry ready to push open the buds. Welsh onion up, tulips, iris. A day lily marching already into the west.

Last day of March, warm and cloudy. Going out, this month, like a soggy lamb, wet from the rain and shivering a little from the chill.

APRIL

April is in, and it will begin with showers, looks like. Soften up the soil, call the worms up, feed the robins. Think I will dance this one, wear the day in motion.

It's nine degrees Celsius out there. Strawberry plants naked of snow, and the furled rowan buds ready to burst out. Heard sun's call. Soil bare in places, wet and dark. Last year's foliage sogged down, mouldy. Revelation of the garden: resurrection. All day the merlins calling overhead, scouting nest sites, speeding toward each other for the dance.

Raining this morning, plus two degrees Celsius, it will turn to wet snow and the streets will be crazy with slush and mushy ice. As if we need more water here, but earth in her wisdom and her needs will give it. Store it, flush out the poisons in the waterways, the rivers and lakes. Humans are not the only beings here who matter.

Sunshine today and the crows flapping busily about. Small birds raid what gets dug up with the plant beds. Everybody is getting ready to nest and make new life. Moving.

Fog, rain and the waters rising. Wet spring. Old stubs of last year's garden still hold soil, compost overflowing from winter kitchen. Last ridges of ice and snow black and melting away.

Another cool day, ice jams still on the rivers, backing up flow, flooding fields. The Red, the Assiniboine, the La Salle, the Fisher, rivers too full from the winter, from spring snow and rain. Water, everywhere, belongs to itself, to the land, to the forests, to the creatures, plants that live here.

First shoots of spinach up, like two hands together pushing through the earth. First little round radish leaves. Grape vines dripping their sweet sap, feeding early bees and wasps. Rowan and hawthorn leaves rolling out, uncurling. Enough perennial onion shoots to snip into a salad. Enough sun and heat to sit on the patio with a glass of wine and friends.

Sun up, crow calling. Bunny passed through the garden, looking for a place to put her young. A little frost on the grass sparkling. Integrity of a world.

Twenty-four degrees Celsius today, and couldn't stay inside. Gardened, took the sketchbook and charcoals out to the patio. Drew the spirits in the hawthorn, faces between worlds. Geese flying low, merlins shrilling their high fast song, crows patrolling. Thought of my children, my grandchildren walking in the sun.

Who knew such a day could call to the spirit. And see, the business of spring goes on in the weather, buds opening, cats prowling, birds skimming the wind. Fierce season.

Birds in the hedge, wind down, and I am listening to the garden breathe. Seeing the faces of spirits in the trees. They are never what you think they will be.

Raining, cloudy, Beltane eve. Everything that electric green of the first rain. Earth dark with water.

MAY

A small bird is singing his little heart out for a sweetie. The hedges have enough leaves now to hide a nest. Comes the rain again. And good sleep. And patches of blue before the next thunder.

Up at dawn to the voice of Fred Wren tweeting his little birdie heart out.

Seedlings in the sunroom window reach east, green as you please; tomatoes, peppers, cosmos, waiting for the heat and frostless nights, waiting for their place in the garden.

Sun shining this morning, oh, sun again, and the green from recent rain. Now give us heat and the gardens will grow like mad.

Drawing, still, faces from the worlds. Green outside from so much rain, and it's raining cankerworms on my roses, they munch their way through buds and leaves, rappelling from the lace-leafed old elm on their thin silk ropes.

Cold today, and rough winds shaking the buds, indeed. Yet they breathe with vigour the spring air, take the slap of cold rain, bury their feet deeper in the opening earth.

Walked the garden early morning: wrens, sparrows, crows, woodpeckers calling. Same as the forest. Dew underfoot, light coming through the trees, into the clearing of the garden.

Watched a small determined merlin harassing several ravens. Little squeaking whirlwind rushing at them, and the ravens answering with their irritated "Piss off" squawks. Birdie dramas.

Wanting rain to retreat, warm to return. Garden full of salad now. It comes on thick and fast and we eat ourselves silly on the greens.

A beautiful Sunday afternoon, light breeze, sunshine. Lawn is mowed, house vacuumed, and I'm drifting now off into the wind, the slight wind, and warmth of the sun. Dragonfly came by to take me on a journey.

Sunny day again, knock wood. Heard the morning birds chirping, trilling and here's me thinking about love, what little we know of it.

Blessed be the earth in her turning, the sun in his rising and setting. Blessed be the dragonflies in their munching of mosquitoes.

THE GREEN DRAGON
WRITING THE BOREAL (SECTION ONE)

How do you write an *oikos*, a home? And how do you find oikos in a living landscape? My first close and intense encounter with the boreal forest came when I was eleven going on twelve years old. My family spent two full months in a trapper's log cabin, across the lake from the mining town of Snow Lake, Manitoba. The full summer. Somewhere there are photos, and somewhere there are paintings, watercolours, of the "cabin in the woods."

And my memory, fifty and some years later, of that forest summer still calls up the evergreen scent of it; the softness of the forest floor, soft from mosses and humus; the crack of twigs underfoot; the lap lap of the lake on the shore stones by our jetty; the nighttime howl of the mine across the lake as it ventilated every evening. Wild berries, wild fish, food from the earth and the waters. I don't recall the voices of wolves, or eagles, or the snuffle of a bear at night while we slept. But I know we heard them. We saw them, and the songbirds. Small ones. I was too young to know or care what they were. But I do remember the jump of fish as someone reeled it in. What the lures looked like flashing in the dark water.

Writing the boreal began a long time ago, so I am starting there and not here. If you listen to an ecosystem and watch it you will learn its economy, its way of ordering itself as a home. Its laws, its ethics, balances. So looking at the boreal as a home – and it is one, an enormous and varied home – how do we read it?

We can read it first from the language origin of all our "eco" words. Oikos: a Greek word, and it means home. The term "ecology" derives from it: "oikos" means home, and "logos" means discourse. Ecology is the study, therefore, of home, and it is everything we say about it. An ecosystem is a home system, an economy.

And this word too, economy, also derives from it: "oikos" again means home, and "nomos" means managing. How a home is managed, then. Or in the case of the forest, how it manages itself. What laws the forest makes and follows in order to flourish. An economy is described by the *Collins* dictionary as "the management of a household and its affairs; careful use of materials. Harmonious organization." To be economical is to be "not wasteful." To economize is to "spend with care and prudence, to use with frugality." The forest, or any left-alone system, already knows how to do this. We are learning, working on it.

Let's say our home is where we live with other beings, usually humans and perhaps a pet or two.

Let's say it contains some houseplants and there is a garden. We have a small oikos then that contains a variety of beings living in relation to each other. But wait – this little oikos does not exist in a vacuum. It exists in a neighbourhood of other small *oikoi*, a community of homes, which becomes a larger home, a larger oikos, with yet more beings – trees, rivers, insects, birds, small mammals, micro-organisms, fish, frogs, snakes. I am speaking here of a city. The city sprawls on a piece of land that is not hanging in the void, but which connects, flows on into larger and larger land masses and communities of beings such as grasses and bulrushes and shrubs and lakes, into forests and tundras and ice caps. Where does our oikos end? It is a whole planet finally, a community of communities, an unimaginably complex set of relationships among beings which both arise from and at the same time shape their environment.

And how do we talk about home? "Home is where the heart is," "a man's home is his castle," "a woman is the heart of the home," "I want to go home," "I want to take you home with me" as though home is the highest and most precious place you can take someone. "Welcome to my home," "I want to go home," "take me home." Home is where we feel safe and home is an expression of who we are.

Home extends outward from our immediate location as the forest, for example, extends toward, enters, the city. The power we use daily is pulled from the waterways of the north, and now the north, the boreal forest, becomes essential to, part of, our immediate oikos, that house with the garden, on the street in a city. And this brings us face to face with the great mega dam projects and the ethical dilemmas posed by this extension.

In 2004, I read a proposal for the construction of a mega dam at Wuskwatim, in Northern Manitoba. Reading it with oikos in mind, with the dilemmas of ethics in mind too, I had some questions:

- A. What is "conservation biology" and what are "natural values"?
- B. Who is this "we" that is doing all this deciding?
- C. Do the forest and its biota have to prove its/their worthiness to be protected? What about humans managing themselves?

The fundamental purpose of this project, a mega dam, is the support of humans, managing the environment for humans, but nothing in the project asks humans to manage themselves properly. Given that humans are not very good at that, how best to "live with" the boreal? Of course, ethically, humans are responsible for their own support; we do have to look after our own species, as does every other species. There is nothing wrong with utilizing the resources of the land, how else do we live? The issue is not that we use resources, but how.

The project (Wuskwatim) would like to "allow" fifty per cent of the boreal to live. How sweet. How very kind. How laudable.

That is, if you are thinking like a conqueror. And one should include Aboriginals in a partnership for decision making. "Include"? What form would that take? And if there is a "partnership" of communities, does it include the communities of the boreal itself?

Is anyone asking the trees? The bears, the bugs, the bogs? In other words, is this ethos entirely anthropocentric? What would happen if First Nations had landowning power, called the "meeting" and "included" other human interests in the decision making?

Suppose the spirit animals and plants were present to listen and consider the requests and needs of the petitioners, and had the power to say yes or no to sustainable development plans? Perhaps what is needed is a legal advocate for the boreal forest itself. Does Nature have rights? Does the boreal forest have rights? Does it have a voice in these discussions? Nature is compassionate and generous, and supports all its biota, including humans, but it is not a carcass to be carved up by the greedy and the hypocritical.

Suppose the spirit animals and plants called the meeting, and asked all humans to attend?

Has anyone asked the forests? For heaven's sake, they don't have to justify themselves by their instrumental value to humans in order to go on living.

Ask the boreal how to live with it, be supported by it and vice versa. No, I am not kidding. We already know what happens when we start blocking animal migration routes and destroying habitats.

Perhaps there is an answer: make the sacredness, the holiness of the forest accessible, in art, in music, in the raising of voices singing. Therefore we created the project now called Boreality: for the development of a piece of sacred music that provided a voice for the forest itself to speak, to sing. We would partner with a Manitoba composer to create a choral piece, authentic sacred music, from, for and about the boreal forest.

And here is our mission statement, our poetics of "green art," if you like, a statement of intent and hope:

Katherine Bitney

The boreal forest worldwide is under threat from human activity including indiscriminate logging, hydro projects, general mismanagement and global warming. This is in part due to a human ethics of utilitarianism and to the desacralization of natural landscapes and ecosystems, and elemental cycles (for example, fires). "Not listening" to the boreal has led to over-logging, clear-cutting and disastrous reforestation through arboreal monocultures. If one sees the interaction of human and forest as a clash of narratives, or as the superimposition of human narratives over forest narratives, one can identify Western humanism as focalizing human need/want/greed and overriding the integrity/needs/autonomy of other life forms and systems, including the boreal forests of the world.

Our underlying philosophy regarding the creation of this composition is the re-recognition of the land and its activities, its own narratives, as sacred in themselves, and we are considering this representation as an exemplar of what we are terming "green art." For us, green art is not simply the use of recyclables for the construction of art, but listening and interpreting/presenting the voices and spirits of the land (here, the boreal forest) itself. This would include seasonal activity, as well as flora and fauna. In a sense, we will ask here that humans enter the narratives of the boreal forest as respectful participants. The natural boreal narrative(s) will alter with human observation and participation, but this will or should alter it as the inclusion of any animal or plant living in the system would, in other words, as part of the natural landscape's own narrative. Development of a boreal choral piece would therefore include time spent by the composers (writer and musical composer) in the forest itself to listen to those narratives. This would include time spent in the forest for all four of the seasons.

The forest makes its own sacred music that is not subject to human text and textualization. Listening without text allows humans to hear the narratives of the boreal without prejudice and without Western humanist focalization. The job of this choral piece is to interpret and re-present those narratives to a human audience.

Boreality began as a collaboration between the Manitoba Chamber Orchestra and *Prairie Fire* magazine, to create the choral piece; an unusual partnership, with an unconventional project, using untested methods. We were flying by the seat of our pants, or blazing a trail, or both. We gathered our team of artists: writer/librettist (myself), composer (Sid Robinovitch), soundscapist (Ken Gregory), photographer (Mandy Malazdrewich) and our coordinator (Janine Tschuncky, who also became, as it turned out, our den mother). We did not know quite how or if it would work, this throwing together of virtual strangers; if we could live together for those short periods; if we could bring our seeing and hearing together into a completed, cohesive piece. But, nothing ventured, nothing gained.

We had originally envisioned an art show of Mandy's boreal photos, with Ken's soundscape as its audio voice, a boreal issue of *Prairie Fire*, all to be launched when the choral piece premiered. Authentic sounds of the forest recorded by Ken were part of the choral piece itself, weaving in and out of the music and the voices raised in song. It was an ambitious multi-year project. Funding cuts would not allow us to do it all, but we did not know that when we set out on our first encounter with the Manitoba boreal forest.

The first year of this project was to be for listening, just listening, letting our ears and our eyes take the forest in. Asking it to speak to us in all its voices: trees, water, wind, animal life, human life, once in each of its seasons.

We began our journey of listening in the winter, close to winter solstice, 2008. We packed ourselves into a rented van one cold winter morning, and headed north to the Hollow Water First Nation to stay with Elder Garry Raven, to listen by his river, then on to the mining town of Bissett.

Winter Boreal Trip, Our First Walk in the Forest, Hollow Water

Winter

We leave the city at 10 a.m., and head north. The roads become narrower and narrower, more and more snaky, into logging country. Passed by logging trucks and little other traffic. We turn off on the gravel roads into the bush. Entering the forest we drive past miles of burned trees, acres destroyed by forest fire, stumps and skeletons and the vigorous heads of new spruce barely visible, but the forest is rising again. Birch and aspen, spruce, pine. Dark blue spruce. Lace of bare birch. Humps of rock.

We are staying first with Elder Garry Raven and his partner, Björk. Garry's land is by a river, white and frozen now, a few buff reeds held in the ice near the banks. Björk feeds us, Garry teaches. His land teaches us also.

In our small cabin at Hollow Water, Sid is reading a Japanese novel, Mandy crochets, Ken plays with his laptop and sound equipment. Janine is reading, writing and talking about food.

They've made the fire and it's crackling in a rusting cast-iron stove covered with stones. Oranges on the table, the den mother's gift. I can see the wide white snake of the frozen river, the bush along its banks.

I've been here before in a winter bush with that brilliant sky tenting lake and forest. Here what you hear is not stars but the vehicles on the road. You see smoke, the cold river, dogs. Crow feet on the snow. The black sky, trees against it, birch, the conifers, spruce pines and nameless mysterious trees.

The doors of our cabin are hung with cedar for protection and cleanliness. Can spirits pass? Why are we here? To listen and listen, learn the medicines of the river and trees. Roots, leaves, flowers, stems. Have our bodies forgotten how to eat?

❧

It is bitterly cold. Was called outside this morning; small birds singing in the high trees, chickadees, redpolls, jays and the raven passing on the wind. It takes a day to shed the city. Trees are very tall and slender here.

We came to see and listen. Spirits. Dogs. The river vista. Sun rising, bare trees open to the colour bands of dawn. Open for the sunrise – it's late, after 8 a.m. this close to winter solstice. Wind up, sun in and out. Pale and lovely. Blue of sky almost not there. There is an edge to sound in extreme cold, a scrinch. Bird tracks, silence. No sign of deer or moose, of wolves or any other predator.

Fire stoked, Ken is off to hunt sound down by the river. Mandy capturing the sunrise, her fingers and her cameras freezing.

Does the forest welcome us?

Must be twenty-five years since I've been in the winter bush, seen sun dogs in the pale blue sky. Annwn, the antechamber to heaven. Noting the sky at night: Orion and Ursa Major, the Wain or Big Dipper. Whole clusters of stars. We forget to live by them. Fire and ice. Sleep and watching.

❧

Thinking on the place and role of the artist here. Is the forest a sacred text? Singing itself. Is a choir like a stand of trees? An orchestra? A river? A land full of medicines?

Sid finds his feet in the cold and in the making of fire. He hears the melodies already. Everybody blooms in the cold.

Coming to Hollow Water is right. We are just starting. Solstice is coming. Sid walks to the bend in the river. Ken is up at the crack of dawn with his microphones, taking the sound of morning birds at the feeder, the far croak of ravens in the woods. Mandy with her camera in plastic, recording the sunrise, playing with the dogs. Janine making twig tea. Land medicine, now there's a book.

Juniper and cedar tea. Root of calamus. Bush food. Garry wants us to do a sweat. Part of the land, the dragon.

Looked out. Sky is dark blue with the coming night. It has started well.

~

We are in Bissett, staying at a motel. Oh the pleasure of a shower! In the evening, we attend a school Christmas concert. So few children here they are called up for their gifts by their first names only. Happy. A beautiful wide sparse Christmas tree with garlands and lights. The wideness of it is reassuring and safe in some familiar way.

This is a small town built on rocks, the Canadian Shield. The houses scattered, lit for the coming holiday. Christmas is all about Santa now, has walked away from its Christian narrative. Not abandoning the sacred, but reaching back, down, to older layers of story about the winter solstice, not quite getting there: elves, Santas, gifts and trees. Something is missing. Once, Santa was Odin. In this deep forest, in the night silence and looking out at the immense darkness, you might almost see the old god riding the sky, the Wild Hunt following him. In the end, though, what we celebrate this time of year is the wobble of the earth as it stops its rolling, tips the north forward again, toward the sun. Is that not in itself sacred?

Garry's partner, Björk, who came from Iceland, had given us a description of the Icelandic practice of scaring children at Jól. Children need to be scared, she told us. And she laughed. Called up the memory of those old stories she carries from her own ancestors encountering the land, the northern climate, the coming and going of light and dark, the old gods she still walks with.

Winter solstice is dark and scary. The glorious sky. A sureness.

~

We went today to the school. Light, movement, love, and so few children. So much knowledge for them. The school has light, airy rooms. Once this school went to grade twelve. The children tell us stories about ice fishing, about bears. Alma tells a story about bears and a tame deer. About mosquitoes, black flies, deer flies. About ice fishing, falling into ice holes. About a bear on the veranda. Alma, who has lived here all her long life, tells us she always walked everywhere, and she still walks

Sid plays piano, and a young girl with braces on her teeth is all agog. Mandy shows how to use a camera. Ken plays with traps and the kids worry he might hurt himself on them. These children learn to shoot bow and arrow, to set traps.

Someone here teaches music. There is a violin, a piano. So few children, such a beauty of a school. The lake is almost idyllic. A slowly dwindling community in a paradisic setting.

We walk the frozen lake, toward lonely islands. The San Antonio mine at the lake edge is forever beeping with its hauling trucks, night and day.

At the school, one child had told of how bright the full moon was once, that you could see the outline of a tree clear across the lake.

We go to the cemetery to find Janine's mother's grave. It is tucked off the road, inside the forest, and the snow is so deep we cannot find the grave. Ken takes soundings, Mandy snaps pictures. We all listen for the scuffle and scurry of animals.

༺

The mine kept me awake all night, and I don't recall any dreams. But I saw landscapes in a half sleep, snow, rocks, river. Dark blue. The solstice is coming. Blue energies. Wolf I see. This is the teaching zone.

DOES NATURE HAVE RIGHTS?
ETHICAL IMPLICATIONS IN ECOLOGY

We ask the question does Nature have rights because we are at an ethical, and survival, crossroads. This question does not need to be asked out of desperation, and for much of human history and prehistory, it has not been. Descriptions of Nature are constructed not just because it is a matter of survival, but also out of awe, wonder and, less laudably, in order to "ethically" facilitate human agendas of colonization. So we have stories of Nature as divine, as subject, inspirited, alive; and we have stories of Nature as dead matter, mindless, object, mechanical. We think of it as with intentionality, and as without; with and without a reason for being; with and without unity; with and without self-determination. Each of these descriptions serves a human purpose, each defines Nature in such a way as to make it comprehensible to us.

Let's begin with some contemporary definitions of rights and Nature. On "rights":

> "That which is consonant with equity or the light of Nature; that which is morally just or due"; "The title or claim to something properly possessed by one or more persons"; "That which justly accrues or falls to anyone; what one may properly claim; one's due."[1]

On "nature":

> "The creative and regulative physical power which is conceived of as operating in the

material world and as the immediate cause of all phenomena"; "more or less definitely personified as a female being"; "the or a state of Nature a) the moral state natural to man, as opposed to a state of grace b) the condition of man before the foundation of organized society"; "The essential qualities or properties of a thing; the inherent and inseparable combination of properties essentially pertaining to anything and giving it its fundamental character."[2]

And another for "nature": "The world, the universe, known and unknown; the power underlying all phenomena in the material world; the innate or essential qualities of a thing; the environment of man."[3]

Nature as power again, and as underlying, being within, the material world. Why is a state of Nature not a state of grace? Where and when did this dualism arise?

The language in these definitions presupposes an anthropocentric, perhaps even Biblical, but certainly Western worldview. But it also presumes an older layer of belief in a creative, organizing power *inherent in Nature itself*, and it recognizes that this power is always personified as female.

Still, these definitions draw a distinction between man and Nature, and between the natural world and the world of God, by using phrases like "as opposed to a state of grace." Yet they describe Nature as "the environment of man," that is, "belonging to." And they tell us that Nature provides "a moral state" for humans. But the language of these definitions also characterizes the world of God (a state of grace) as "other" than Nature, beyond, above or injected into Nature, from the realm of the divine.

With regard to rights, Nature again plays an ethical role – "that which is consonant with equity or the light of Nature" – as though equity exists "with the light of Nature," inherent in it. But while equity is not quite equality, it is a characteristic of Nature.

Can we call this natural equity? Is Nature the source of ethics? Of rights, including its irrefutable essence, its character beyond any interpretation? The problem arises always with how that essentialness is constructed (and it is always constructed), by whom and toward what agenda. With regard to the question of the rights of Nature, one must ask the question first, what is the Nature of Nature? Even, does it have a Nature? Could an account of Nature that would have rights be constructed without essentializing? Anthropomorphizing? And if Nature has rights, what would they be? If, as the dictionary definitions suggest, only persons have rights, can we, in reassigning rights to Nature, avoid personifying? Do we need to? There are myriad accounts of Nature; how do we choose?

While much has been made of the negative effect of the Biblical account of creation and the God-Nature relationship in criticism on ecological issues (and with justification), not enough, I think, has been said of the role of Greek thought in Western discourse about Nature in the past couple of millennia, and its *negative* effect on the relationship between humans and the rest of the natural world. Val Plumwood identifies "the Greek" as a wellspring for the Western ecological dilemma, in her book *Feminism and the Mastery of Nature*:

> The society of classical Greece is often viewed benignly, by both liberal and environmental writers, as the cradle of western civilisation, and the philosophy of Plato is especially revered as the source of its proud intellectual, artistic and civic traditions.... [But some feminists] have seen in the Platonic account of reason a masculine identity which has profoundly influenced its character.[4]

She points out that many environmental writers look to the Greeks as "respecting and celebrating the earth through the worship of Gaia," and this may have been true on a country level, but as she demonstrates, even the Gaia story in the hands of Plato is "designed to promote...not environmentalism but militarism."[5] This is because for Plato, as for other Greek writers of the classical period, the focus of ethics and myth is not Gaia in the old sense, Earth, Nature, but polis, city. So it is perhaps with some irony that, in their collaborative work in the 1970s, two well-known Western scientists, James Lovelock and Lynn Margulis, created a scientific model for environmentalism positing the earth as a living, self-regulating organism, which they named the Gaia Hypothesis, for the ancient Greek earth goddess.

I argue that not only does Nature have rights, but that they were systematically stripped away by redefining Nature, from living being to woman to slave to mechanism. Nature became an object, non-agentive, non-living; in the Greek sense, non-logos. According to Lovelock, "The idea that the Earth is alive in a limited sense is probably as old as humankind."[6] Undoubtedly. Even today, this belief persists elsewhere and in other cultures: not only in societies with animist spiritual traditions, but also in Hinduism, Buddhism, Jainism, Taoism, some of the great world religions. How, then, and why did Nature become a slave, an object, a non-living mechanism without identity or rights? How did Nature become not us?

As the dictionary has told us, and as we already know, where Nature is personified, it is always personified as female, as mother. So maybe a good place to begin is where these two together, mother and Nature, are publicly stripped of personhood and of rights. I offer as signifier the following excerpts from the play *The Eumenides* by Greek playwright Aeschylus. *The Eumenides* (meaning, more or less, nice old ladies) is the third play of a trilogy on the Trojan War, called the Oresteia. It marks a point, not only of what Luce Irigaray has identified as the suppression of female genealogy,[7] but more significantly for this study, a point at which mother ceases, not only to have rights, but to be.

This Greek soap opera tells the story of how Orestes, son of Agamemnon, kills his mother because she killed his father because he had sacrificed their daughter Iphigenia for a fair wind to Troy. The ghost of Clytemnestra, the murdered mother, rouses the Furies, the ancient retributive arm of Gaia, to bring him to justice. Notice that justice, and therefore ethics, was assumed to arise from Earth, Nature. The solar Olympian deity, Apollo, who admits to having incited Orestes to kill his mother, argues for Orestes in court:

> The mother, so-called, is not the child's begetter,
> but only nurse of the new-sown embryo; the one
> who mounts, the male, engenders, whereas *she*,
> unrelated,
> merely preserves the shoot
> for one unrelated to her.[8]

As proof, he offers the goddess Athena (who, we will all remember, was born full-grown from Zeus' forehead):

> a father could give birth without a mother; near
>   to hand
> there is one who was not nurtured in a womb of
>   darkness
> but is the kind of shoot that no goddess could
>   give birth to.[9]

In this short speech Aeschylus/Apollo reduces woman to body/field/Nature and strips away her rights as mother, indeed, strips away the word and meaning of "mother." In this speech he not only disconnects child from mother, but human from Nature, from the Earth itself. Here, Earth is a field where the seed grows, but with no life of its own. The Furies are argued out of supporting mother-right by a soothing speech from Athena, who is made by Aeschylus to repudiate her own mother (Metis, who was actually swallowed by Zeus before Athena's birth out of fear of the child she was carrying), and to announce that "there is no mother who gave birth to me / and I approve the male principle in all things and with all my heart."[10] She casts her vote such that Orestes is acquitted of matricide because he had no mother to murder in the first place. It is not accidental that Athena is the goddess of war and is identified with polis, not wild "female" Nature. Thus she is herself stripped of her femaleness (and her older, chthonic identity) to act as poster girl for the glories of patriarchy. Athena – who was once Atana, Tana, Tanit, Danu, ancient earth mothers, snake goddesses, wielders of natural justice, and oh, Medusa, whose fierce serpented head she carries now on her shield – had her own wild nature tamed.

The full implications of this story still resonate with us. No, we do not repudiate mother, but we do repudiate Mother Nature. And where we acknowledge her, we are at war with her, find her vengeful, furious and so on. Or, we try to impose a "peaceable kingdom" model on her from the father on high, the external god. More importantly, the shift here from mother-right to full-blown patriarchy provides an ethos for domination – of male over female, Greek over barbarian, master over slave, of human over Nature.

It provides ethical and legal justifications for war, slavery and colonialisms. It unpersons woman and Nature simultaneously, disempowers both, constructs them both as "mere," "just" and "only." It is only a short hop from the not-us-ness of women to the not-us-ness of other humans who may be wanted for slave labour. Exactly what biology Aeschylus is using to justify the abolition of "mother" is not altogether clear, but with the writing of the Oresteia we are in the fifth century BCE already, and the processes of human anatomy are by this time well known. Mother and Nature can no longer legitimately defend themselves, even if they still had their rights, because their retributive arm, the

Furies, has been neutralized. In this story the human female has been stripped of her rights, her personhood, her humanity and re-inscribed as non-human or less than fully human: she is a body, something to be ploughed, used and discarded, just like Nature itself. It is evident in this account by Aeschylus that the rape and plunder of Nature is no sin, since it has no more rights, is no more a person, than the human female.

So in these few lines we move from one set of guiding beliefs to another. This rejection of the "womb of darkness," the feminine, is a distancing of the human from identification with the body and Nature, and from the ethos of the natural world – just as the Delphic oracle's original source of knowledge and ethics (Gaia, Earth, Nature) is forcibly replaced by the violent inhabitation of the god Apollo. Indeed he literally murders the old order, the Python, and rides the oracle herself like a rapist.[11]

Here the voice of Nature, of Gaia, the Earth, as a source of knowledge and ethics is displaced by a deity who anoints himself with the right to speak, to articulate prophecy, through violence. Might makes right(s). Thus Earth and woman, who were speaking subjects, are silenced and objectified, and become merely vessels through which the male voice passes (witness the bizarre speech of Athena). By the male gods' appropriation of such powers over the female, "'Mother' Nature is depotentiated, and her birth-giving powers are taken over by the male."[12] And depotentiating Nature rationalizes the eventual wanton destruction of Nature we are even now scrambling to undo. Aeschylus' unmothering here is heard all the way down to Freud's strange (or in this instance, not so strange) belief that the human (male) is forever trying to repudiate, distance himself from and simultaneously appropriate the mother.

Aeschylus' text opens two main areas of discussion: first, of course, the simultaneous description of woman and Nature as identical (and as collectively other), and the construction of the patriarchal ethos as right, justified and divinely sanctioned.

Secondly, the revelation that everything from culture to ethics can be, and is, constructed, and as such can be de- and reconstructed. The nature of Nature and the nature of woman, in this text, as in countless others, are fully identified as one and the same, and must, I think, be de- and reconstructed together.

I will stipulate that feminism is ethics, in that it begins from a positive ethical position: that gender equality is right, and naturally right, and that all forms of inequality and domination are wrong. By this yardstick, every exercise of power over someone that disadvantages the receiver is morally wrong. For the receiver to be disadvantaged they have to be able to experience disadvantage, that is, they need to be alive and feeling. In order for the disadvantaging to be justified, the receiver has first to be turned into an object, into an "other" that does not inherently have the same Nature or essence as the exerciser of power.

At last count there are 260 billion galaxies in the known universe. This is Nature. These galaxies are filled with planets, stars, nebulae and dust, energy and movement, heat, light, fire. When we speak of Nature, we don't always take these awesome facts into account. When we speak of Nature we speak in particular of one small planet. But that planet is our home. And while I do not wish to appear parochial, nevertheless I wish to identify planet Earth as the primary locus and focus of this discussion of Nature. That we humans have already reached beyond our own biosphere does not mitigate this identification. Yes, we have put space junk on the moon, on Mars and even sent a satellite spinning out somewhere beyond our solar system. But what we live in daily as Nature is the flora and fauna, the seasons, the phenomena of Earth. And the immediate ethical problem of our relationship with Nature is with our own planet.

Now I do not forget here that Nature is not just the matter of the universe, but is its movement also. It has power, or energy, and habits of continual change, creativity, evolution. Earth, and

Nature in general, is a system of systems which we do not fully comprehend, neither in its behaviours nor in its, dare I say, Nature.

It is quite literally our home, our oikos, and, Aeschylus notwithstanding, we need to remind ourselves that it is indeed our parent. We humans are made out of it. It gave birth to us in whatever way a planet gives birth to its life forms. With this in mind, how can we pay scientific lip service to this truth and yet at the same time imagine ourselves as other, or, more properly, imagine the planet as other than us? The Greek story cited above, like its Biblical counterpart, identifies the source of human life (complete with that strange factor, the mind) as a sky father god. These fathers absurdly try to give birth or create matter without the matter/mother having any value or ontological status of their own beyond their instrumental function to the father. In the Greek account, Nature/woman is stripped of rights. In the Biblical account, Nature has no rights of its own to begin with. It is the property of God. Both accounts are violent in that sense. Both accounts objectify. This is the official legacy of Western thought.

There is of course the unofficial legacy, what Irigaray calls "a sedimentation laid down in its time by earlier traditions."[13] She is referring to the older beliefs of the inspiritedness of the natural world, what is nowadays being termed "The Goddess Tradition." Irigaray goes so far as to suggest that the death of the phallic god (the one who objectified Nature and disinherited women) heralded by Nietzsche and company is "not about the disappearance of the gods but about the approach or the annunciation of another parousia [coming] of the divine."[14] An optimistic thought, and not beyond imagination. Perhaps this is Lovelock's Gaia? This sedimentation is the shadow tradition of Western thinking, it appears as a St. Francis of Assisi, or as a Hildegarde, or as a Keats, or a Thoreau, or a Spinoza, again and again. These are marginalized thinkers: poets, mystics and heretics. But there it is nonetheless, a persistent source for the reconstruction of the rights of Nature. Yet we must first deal with the legacy of estrangement from body and Nature that we have inherited from our Greek and Hebrew forebears: a system of dualisms which has left humans excluding each other, other species and the planet itself. A system Plumwood describes as "hyperseparation," where we (all of these "others") have been forcibly marked as discontinuous from each other. Nature itself, Plumwood argues,

> in most of its senses and contrasts is subject to *radical exclusion*, and is conceptually constituted by it, as well as by the other features of dualism....
> ....[Therefore,] there is a total break or discontinuity between humans and nature, such that humans are completely different from everything else in nature.[15]

Different and superior, that is, masters of one or another ontology of Nature which leaves it in the "slave" category. "That Nature has been constructed as a means to human ends is not news, nor is it news that the Western view of Nature does indeed categorize it as slave. But it is a disobedient slave, just like woman, and without benefit of phallic rule, it is characterized as lacking, just as the Lacanian/Freudian/Aristotelian woman is lacking."[16] As Plumwood points out,

> that women's inclusion in the sphere of Nature has been a major tool in their oppression emerges clearly from a glance at traditional sources: 'Woman is a violent and uncontrolled animal' (Cato 1989:193); 'A woman is but an animal and an animal not of the highest order' (Burke 1989:187); 'I cannot conceive of you to be human creatures, but a sort of species hardly a degree above a monkey' (Swift 1989: 191); ...

> 'A necessary object, woman, who is needed to preserve the species or to provide food and drink' (Aquinas 1989:183).[17]

Just like Nature itself, no? I repeat these quotations as a reminder that the correlation of woman and Nature, with both described as less than men, is not only not new, but is continuous throughout Western patriarchal history. The disdain for natural and animal life evident in these statements prevents Nature, animal, woman from entering the sphere of those who deserve rights. Given then that the construction of anything or anyone not a human male (and a white middle- or upper-class male, at that) as not deserving of rights is old and constant through European history, is it any wonder that we are at a loss as to how to address the issue of the rights of Nature when we have only just begun to address the rights of women and other "others"?

One approach that feminist writers such as Plumwood, Irigaray and Grosz have taken is to reimagine the terms "difference," and "other." Difference, they have argued, is not a matter of hierarchy but of Nature.

Plumwood, for example, has offered "a notion of otherness as *non-hierarchical difference*. The resulting concept of relevant otherness avoids exclusion,"[18] the kind of exclusion that makes the other inferior. But,

> recognition of and respect for the intrinsic value of the other is an essential adjunct to an ethic of care and respect for difference.... To the extent that respect is directed to the other for its own sake, it will not be respect just for those aspects of the other in which it resembles us, and hence will entail respect for difference.[19]

She also offers a reading of other which she presents as inclusive, or rather as included, but which does not reject difference: she identifies other species, and individuals of other species, as "earth others." "We can relate to earth others in terms of mutual exchange and transformation, and 'the dance of interaction.'"[20]

Her description of earth other could turn out badly, as it still marks beings as others, still recognizes otherness. However, her attempt here is to construct a relational model that recognizes difference without hierarchy, without speciesism, as it were. The term "intentional," which she uses to refer to the earth other, is deliberately chosen to propose an equality of aliveness of other species, not only among themselves, but in relation to the human species. Thus, "the ecological self can be viewed as a type of relational self, which includes the goal of the flourishing of earth others and the earth community among its own primary ends, and hence respects or cares for these others for their own sake."[21]

So a series of interrelated individual interpretations emerges, where beings are considered to be autonomous and independent as part of Plumwood's account of the natural world.

Her ethics also recognizes common origins for all, but allows for differences in "agency," and states that "making room" for the earth other as equal means recognizing and accepting limits to our own agency. This she calls "mutuality." And, rather bravely, she also suggests reassigning to Nature, and to earth others, such mind qualities as intentionality, emotion, agency and teleology.[22]

Nature itself may be seen as being goal-oriented! This brings us very close to an account of Nature as alive and thinking, a subject. For our purposes, to determine if Nature has rights, Plumwood offers arguments that lead straight to the point. She urges us to see earth others as "other nations," as some Native cultures do, and such an account of other beings as equal but with some differences is close to what Plumwood is looking for. She has built up the bones of an ethics of natural rights for all beings, but does not go so far as to fully develop a being for Nature itself: rather she describes a web, a relationalism among beings that constitutes a sort of whole, which we might call Nature. Its nature is collective, rather than singular, and I believe that, like many other thinkers, she is still not free of the dualism of the "one" and the "many," that is, she has not recognized that they are not mutually exclusive. But do we need a one? For Plumwood, the nature of Nature is as a sort of living, intentional, teleological, self-determining and feeling collective of related and relational beings, having, and being a part of, a mutual source: the Earth. This is a kind of whole, though she is very careful not to homogenize and obliterate individuality and difference.

Feminist thinkers are, of course, not the only ones to have attempted to construct an account of Nature that would be eligible for rights. There is also the Gaia Hypothesis, which comes to us not from philosophy, but from science. It proposes that the Earth is a living, self-regulating organism. As an originator of the concept, James Lovelock writes, "the idea that the Earth is alive is at the outer bounds of scientific credibility."[23]

But it is not a new idea, which is why it was named after Gaia, whom we met in the story of Orestes. And not new in science, as Lovelock tells us: in 1785, Scottish scientist James Hutton proposed to the Royal Society of Edinburgh "that the Earth was a superorganism and that its proper study should be physiology."[24] A living body, if not one with a mind. As a superorganism it is at least in some way organized. Lewin reports that "today, researchers are viewing Gaia and the whole notion of the superorganism in the light of the modern mathematical theory of complexity."[25]

A superorganism "can be thought of as a group of individual organisms whose collective behaviour leads to group-level functions that resemble the behaviour of a single organism."[26] Here is an excellent description of the one and the many! This theory, which recognizes *co-operation* as being equally important as the struggle of the individual organism for existence, is very similar to the Gaia Hypothesis.

Lovelock is clear about the ethical implications of Gaia: "In Gaia we are just another species, neither the owners nor the stewards of this planet. Our future depends much more upon a right relationship with Gaia than with the never-ending drama of human interest."[27] I don't suppose Lovelock noticed the discrepancy in that statement: no self-interest in the interests of self-interest! And what is a "right relationship" with the Earth? Or with Gaia? How do you behave toward a goddess?

Eliot Deutsch argues that "natural reverence" is essential to the development of ethical behaviour toward the planet, to Nature. "Without what I am calling 'natural reverence' I don't see how it is possible for us to do more that work out temporary, makeshift adjustments in our actual working relations with our natural environment."[28] He argues for a "metaphysical grounding" for the development of "natural reverence," and one is hard pressed to see how it could be done otherwise. While not letting the East off the ecological hook, Deutsch does recognize "that a great deal of the ecological crisis stems from Western-based technology and the metaphysical and axiological positions that have sustained and nurtured that technology."[29]

A shift in metaphysics seems unavoidable, and Deutsch suggests borrowing from Eastern sources. I will not disagree with this suggestion, but will also not pursue this avenue in this essay. But on the important point of metaphysics: with Plumwood's account of Nature, we already have a nascent metaphysics of Nature on which to build. Like Plumwood, Callicott also describes a kind of relational self as vital to a metaphysics of ecology.[30] While much postmodern criticism wants to do away with metaphysics, these writers seem to assume that metaphysics is a tool, not an end in itself, one that is mutable, and shifting, much like the movement of natural energies. "In the 'organic' concept of Nature implied by the New Ecology, as in that implied by the New Physics, energy seems to be a more fundamental and primitive reality than are material objects… An individual organism, like an elementary particle is, as it were, a momentary configuration, a local perturbation, in an energy flux or 'field.'"[31]

This holistic account of Nature squares with Gaia, with superorganism and with Plumwood's community of interdependent "earth others." Like her model, it recognizes difference and individuality without isolating or privileging them. Like Plumwood, Callicott calls for the abolition of the human ego and champions a metaphysics of continuity between humans and the rest of the natural world.

I have not described all current accounts of Nature, but suffice it that there is no shortage of conflicting stories and descriptions. And suffice it that there is no shortage of conflicting views on what rights are and who or what should have them. Does a river have the right not to be polluted? Does the tiger have the right not to be hunted to extinction? Does the planet Mars have a right not to be terraformed by colonizing humans? Do plants have the right not to be eaten by ungulates? No, I do not believe these are silly or idle questions. If we are to develop an ethics of natural rights, all aspects of Nature must be considered. True, such questions can begin to fragment Nature, to individualize such that if we answer yes to the question of do plants have the right not to be eaten by ungulates, then we imply that ungulates have no right to eat, therefore to live, and so on. In that sense, the community or field of Nature must be considered first, before the rights of individual species, or an individual member of a species. If the rule "everything has a right to live," for example, were to be rigorously applied to all beings, everything would die from lack of food. So perhaps one should ask not only what rights does Nature have, but what rights does Nature give?

The feminist description of ecological ethics given above wants to construct a metaphysics of Nature which would account for the rights Nature gives its members: the right of all beings not to be dominated by others, but to live in respectful relation with each other. Irigaray suggests that ethics begin in wonder.

In *Posts: Re Addressing the Ethical*, Dawne McCance counsels searching among the marginalized, the excluded, for sites for the development of ethics. Plumwood begins with the excluded (women, other races, other species, etc.) and opens her ethics to all.

Returning to the question does Nature have rights?: Plumwood's account of Nature tells us that yes, Nature is being, if not *a* being; Lovelock and Margulis tell us Earth (if not Nature as a whole) is a living being. Who or what, then, has rights? Does the superorganism have rights to super-organize in its own way without interference, for example? Historically, rights have descended from self-privileging upper-class males to males of other classes to females to persons of other races to animals, and now, finally, to Nature itself.[32] So, women are accorded personhood and get the vote, other races are recategorized as equally human, animal is recategorized as living being with feelings, and so on. In each case, the "nature" of the being is redefined.

We could say, then, that facets of Nature are slowly being given rights, bit by bit, according to a fragmented schedule of individualism and hierarchy. Because accounts of Nature (including scientific accounts) are so diverse, and because metaphysics is out of fashion (not to mention a little bit dangerous), it is very difficult to call for the rights of a Nature that is a being-in-itself. Yet one wants to. One wants to apply justice to the murderers of the planet, to be Nature's advocate in court. Perhaps we could bring back the Furies?

Something instinctual tells us that trashing Nature is wrong, is inherently unethical, as even Aeschylus surely knew that abolishing mother-right was wrong: it had to be done violently, and by arguing against Nature, against the natural order. It required the construction of both a rationalized, false biology (women are not related to their children), and a revised metaphysics of Nature, as instrument rather than as agentive being.

Oppressed groups have given an account of what it is like to be oppressed. It is these accounts of oppression that are altering our ethics toward fellow human beings. How does Nature give an account to us of the effects of unchecked human colonization and exploitation? If we understand the Earth as a superorganism, can we deduce its suffering, as we would deduce suffering from, say, a flatworm? A colony of bees? What methods can we use to hear the voice of the being, Nature? If we are a part of the superorganism Nature, then are we able to consider ourselves a part that also suffers?

In response to the above, and other, questions, I am proposing that:

A. while ecological devastation is not by any means confined to the West, it has, as Deutsch has suggested, its metaphysical and ethical origins in the West;

B. these origins are inscribed in and by the dualisms and otherings of Western patriarchal theologies, philosophies, sciences, and agendas of conquest, domination and colonization;

C. these agendas have constructed ontologies of the natural world, which justify the instrumentalisation of Nature, along with all designated others;

D. while other cultures offer models of the natural world suitable to the development of environmental ethics, the West must examine, critique and restructure, and where necessary rewrite, its own myths, theologies, philosophies and ethics first, in order to undo the collective internalization of what I call false values, and to prevent further export of these values.

While there is in the West a lively tradition of environmentalism, and while it has made some useful inroads, nevertheless, until environmental values and ethics are internalized, they will not be implemented to any appreciable effect. The West must make increased use of its laudable tradition of human (and to a limited degree, other animal) rights and open the concept of rights to other organisms (including superorganisms). Those cultures where there is already a description of Nature as alive and sentient should internalize this description and implement laws based on ethics pertaining to living beings. But I will also note that, as feminist critics point out, any culture developing ethics of equality, or rights for Nature, must first look to their own social structure. It does no good to apply rights to Nature if women are still colonized by men or lower castes by the upper castes, where there is slavery or racism, or where species of animals are poached to extinction for profit, and where theologies care more about preserving themselves than about the effects of human overpopulation. Creating hierarchies of rights (more for me, less for you) does not square with an account of Nature as a web or as a community.

But if we are to build a metaphysics of Nature in accordance with these values and these descriptions, how would this account jive with Western theological accounts of Nature as "God's handiwork," "God's creature," "God's property?" We cannot unwrite the Bible, nor the Qur'an. As Lovelock has said, we are not stewards of Nature, we are part of the superorganism, what he has called Gaia. We are just another species.

This scientific description of Nature (as Gaia, as superorganism) goes beyond science and into theology: it is a serious challenge to the received wisdom of the external, all-powerful king deity. Any description of Nature as alive, autonomous, subjective and especially as self-determining (as we have in Plumwood's, Lovelock's and other accounts of the phenomenal world) is on a collision course with Western theology.

I don't have an answer to this dilemma. It may indeed be, as Irigaray has proposed, that we are looking toward another incarnation of the divine, coming out of science, out of feminisms and anti-racisms, and out of environmentalisms. A "messiah" of ecology may be what people are able to hear, yet another mouthpiece for the Voice from on High. (Maybe if we are lucky God will announce his retirement as Master of the World, and declare all mastery to be sin. Can God say, Gee, I'm sorry, I was wrong?) But I must ask, are we still so adolescent that we can only listen to gurus and messiahs? Must we still play follow the leader? Couldn't we just learn to listen to the voice of Nature itself again? I suggested earlier we might want to bring back the Furies. But I wonder if Nature is not already in the process of doing just that.

Notes

1. *The Compact Edition of the Oxford English Dictionary,* s.v. "right."
2. Ibid., s.v. "nature."
3. *Collins Universal Dictionary,* s.v. "nature."
4. Val Plumwood, *Feminism and the Mastery of Nature* (London: Routledge, 1993), 72.
5. Ibid., 73.
6. James Lovelock, *The Ages of Gaia: A Biography of Our Living Earth* (New York: Commonwealth Fund Book Program, 1988), 10.
7. Luce Irigaray, *An Ethics of Sexual Difference,* trans. Carolyn Burke and Gillian C. Gill (Ithaca, NY: Cornell University, 1993).
8. Aeschylus, *Eumenides,* trans. A. J. Podlecki (Warminster, UK: Aris & Phillips, 1989), 103.
9. Ibid., 103.
10. Ibid., 109.
11. Norma Lorre Goodrich, *Priestesses* (New York: Perennial, 1990).
12. Anne Baring and Jules Cashford, *The Myth of the Goddess: Evolution of an Image* (London: Penguin, 1991), 335.
13. Irigaray, *Ethics of Sexual Difference,* 115.
14. Ibid., 140.
15. Plumwood, *Feminism and the Mastery of Nature,* 70.
16. Ibid., 53.
17. Ibid., 19.
18. Ibid., 58.
19. Ibid., 160.

20 Ibid., 156.
21 Ibid., 154.
22 Ibid., 135.
23 Lovelock, *Ages of Gaia*, 3.
24 Ibid., 10.
25 Roger Lewin, "All for One, One for All: Damned for Decades as Quaint and Wrong-headed, the Notion of the Superorganism is Being Reinterpreted in the Light of Modern Complexity Theory," *New Scientist*, December 14, 1996, 28.
26 Ibid., 28.
27 Lovelock, *Ages of Gaia*, 28.
28 Eliot Deutsch, "A Metaphysical Grounding for Natural Reverence: East-West," in *Nature in Asian Traditions of Thought: Essays in Environmental Philosophy*, ed. J. Baird Callicott and Roger T. Ames (Albany, NY: State University of New York Press, 1989), 259.
29 Ibid., 260.
30 J. Baird Callicott, "The Metaphysical Implications of Ecology," in *Nature in Asian Traditions of Thought: Essays in Environmental Philosophy*, ed. J. Baird Callicott and Roger T. Ames (Albany, NY: State University of New York Press, 1989), 63.
31 Ibid., 59.
32 Roderick Frazier Nash, *The Rights of Nature: A History of Environmental Ethics* (Madison, WI: University of Wisconsin Press, 1989).

Summer

JUNE

Cool day, high wind, clouds chase sun chases clouds. Smatter of rain on the windows. A wren, a crow. Voice of Raven, passing. You could lay your head in the lap of her call like it was sleep, safe dream.

Looking out my study window at the little lopsided oak we planted on the boulevard maybe seven years ago. How slowly they grow, scrub oaks. This one has endured, survived, winter after winter of being buried in heavy banks of snow by the ploughs. All plant life impossibly green now from the constant rains, and the little oak too. Its beautiful leaves green as you please, waving in the wind.

Sunshine today, and the crows still flapping busily about. Small birds raiding what gets dug up with the plant beds. Everybody is getting ready to nest and make new life. Moving.

Wind picking up, air cooling. Rain coming again. Lucky, lucky, we had a morning of sun to get the jungle mowed. Few days and it will be deep again, perennials marching into the lawn bold as a flock of grackles. Honeysuckle vine now opening its thin orange-and-yellow flutes, its big bobbing blooms like birds of paradise.

Fred and Ethel Wren toing and froing, feeding the chicks in the wren box. Fred scolds from the fence if we get too near the rowan tree where we hung their house this year. Tough little bugger, Fred, with the sweetest voice in the garden. He will stop singing when the chicks are fledged, and I will miss his song, greatly.

Pouring rain out, dashing petals to the ground. Spirea, roses make puddles of pink on the grass, sidewalk. Wasps trying to escape into the house. So be it. This is summer, already.

Cuthbert Grant rose now open, deep crimson, velvet. High scent of the true garden rose. Oh joy!

Bees going bonkers in the open roses, hovering expectantly around the just-opening False Indigo. Bees, bees, the big bass bumblebees, the leafcutters, honeybees from somewhere, the ground bees. Such music.

Cool evening, wind picking up. Rain again. Maybe thunderstorms tonight. Earth is saturated and still the rain comes. Wrens calling to the young to come out of the wren box, even in this late day drizzle. Time to fledge, and mother is chittering for them to come.

Looking at the boreal forest, the city, an edge to the forest, an entrance.

Wrens have moved their children out of the wren house, hiding them in the bushes, teaching them how to fend for themselves.

Oh the garden. Climbing roses in their pinks and magentas, on fire with their ecstatic blooming. Perennial poppy opening its first great orange hand of a flower, more to come. Slender flag irises, blue and delicate on their tall stalks. All open, all calling to the sun to feed them.

Work done. A laden breeze seems to carry the past in it as sometimes summer winds do.

Clouding over, rain coming again. Rain every day, every day. Crazy weather. Who ordered it, anyway?

William Baffin climbing rose now ablaze with bright pink blooms by the back door, would knock your socks off to chance upon it. John Cabot climber by the front, open and singing with bees. Oh how the garden hums when things are in full bloom.

Wild house. Baby bunny hidden in the astilbes. New "child" now the wrens have fledged.

Fourteen degrees Celsius, grey sky this solstice morning. Fred Wren singing, Ethel Wren calling with her purr. Every rose bush, cane in full bloom. Strawberries to pick, coming thick and fast. Some for us, some for the birds, squirrels. Bees hanging upside down on the honeysuckle. Deep green, this world, deep green.

Passing the summer afternoon in the garden with the drum and the sun. And a glass of red wine. And the dragonflies.

JULY

About summer and the thickness of air, quality of memory, mostly youth.

Here come the hollyhocks, suddenly, everywhere. Asiatic lilies, martagons, every colour, everywhere. Rudbeckia, yarrow, bergamots red and pink, delphiniums of every blue. Dark red suns of gaillardia, black hole centres. Hydrangeas. Coralbells. Go walk the streets.

Out of the morning fog I hear the voice of Raven, nearby. I am distracted by the bees. Squirrel is eating strawberries. The crows fly over. Something is missing.

Remembering the summer forest, the carpet of wild strawberries, the songbirds, the door opening and deer visiting. Sound of float planes, the high waters of the creeks. Bugs bugs bugs. And drumming by the water.

An hour ago, I had a monarch caterpillar in J form. Now there is a pale green chrysalis. Still wiggling about in there, shifting, getting ready for transformation.

Cool morning, clear sky. Respite from the heat, for a day or two. Already, time to make jellies from thyme, mint. Peas coming on thick and fast. Too much lettuce. And the lilies opening one by one. Days of beauty and abundance.

Heat wave passing through, up into the thirties, and we are begging for rain, thunderstorms, a break in the air tension, pressure, humidity. Everything bolting that can, in the garden. Lettuce, spinach, sorrel. Birds stay in the hedges, the trees. Squirrels run in the morning while it is cool. Who can blame them?

It's thirty degrees Celsius; walked the neighbourhood anyway. The big elms hold their long wide sleeves over the streets. Roses fading now in the gardens. But oh, the day lilies are open, gold cups on long thin stalks, swaying just a little, just enough. Lavatera, cornflowers, poppies. River shines in the evening sun, walking its bed with leisure, smiling. You think, July. Climax.

Heading up to thirty-four degrees Celsius today, no rain in sight. Watering the garden now, earth cracking in the heat. Birds moving in the cool of morning. Wrens in the bushes, crows on the wires, in the tall trees.

Dark clouds, hot morning. Calling for rain, rain, our hearts' desire, a good boomer to break the air open, soften the land.

Three new monarchs, clustered together on a drying piece of milkweed, their new full wings folded, waiting for the day to warm them into flight.

Who has seen real wild prairie? Yes, pretty much gone. We now imagine the prairie as all wheat fields, crops in gigantic squares. The rest is roads. Can't walk it anymore. How do we listen to the land, then?

Feel like sleeping again. Sleeping in the wind. The ravens have gone afield. Fall coming. Wren babies growing. Joy of the raven's voice, the forest birds are passing through. City becomes forest. Food for the journey south.

Red squirrel running the fence, his highway. Little red-headed sparrow follows, harassing. Two house sparrows rush out of the grape thatch. Bees in the grass, sucking on clover.

This time of year, shifts and changes. Rowan berries start coming in yellow. Virginia creeper shows its first red leaves. Flocks of sparrows invade the garden, fly in gangs. Bees intensify their gathering. Wasps begin to hunt in earnest. Sun angles deeper south, fills the day air with a deeper heat. Beans, cucumbers come ripe, want harvesting. Dill heads up, coriander turns to seed.

All work done for the day. Garden, cooking, book. Walked in the deep wind of summer, spirits riding it, what it carries of animal, plant lives, full and dreaming. Eyes closed in the day's heat, in the warm nights.

August eve and we are harvesting. Herb bread made for feasting Lughnasadh, turning the wheel again. Moving into the velvet air of August.

AUGUST

Odin's crows, one with his back to you. Thought and memory. Who am I without the hunger?

Garden is a jungle of greens, bees, butterflies, dragonflies, summer birds. Decay and climax in August, harvesting in earnest. So much to pick and process and serve up for dinner.

All gathered in for the night. Children asleep and the forest goes on breathing. Will I dream of it tonight?

Sun has come back, bringing such a light to late summer. Harvest and process again today, and consider the role of the stars in timekeeping.

Saw a sphinx moth hovering and feeding like a hummingbird from the night flowers.

Yes, the role of stars in timekeeping. We have forgotten this. Gotten a bit parochial with just the sun and the moon, and lazy with charts and clocks. Yet the Perseids fall like clockwork mid-August, Orion hangs close and low in winter.

Beans to freeze, beets to store. Grapes turning now with their sheen of ripening. High corn making cobs. Monkshood heavy with dark blue blossoms. Garden in August. Beets to freeze for Yule dinner. Pick and dry some herbs. Store cabbage. And in the back of the garden, pumpkin plants running amok.

More heat coming. Sky glowering this morning. Rain perhaps. More processing of garden. In August, it comes on thick and fast.

Raining steadily into late evening. Poor worm out in the wet, sliding along the sidewalk trying not to drown. Me sitting out again, watching the night.

Sun shooting light across the street, the garden, finally, now it is coming to evening.

Felt fall in the wind today. Corn ready and sweet. Onions curing. Time to pick the little grapes. Waiting now for the tomatoes to ripen.

And again, a rainy day ahead. As if the earth needs to soak up well before fall and winter. So green still, a few yellow-leafed twigs dropping to the grass. And the wasps and bees more urgent in their gathering.

Rain garden calls. There are weeds to pull, muck to slip in, peas to pick, doorways to stand in, half in and half out of the rain. Pink on the underside of the clouds this morning as the sun rises. Again promising rain. Restless night, dreams of anger and yearning, urgency. Can't remember a single one. Is someone, something calling?

Coming off the garden today: basket full of cukes, basket full of beans, four cabbages, basket full of tomatoes. Freezing and pickling tomorrow, oh yes.

Minutiae of one's day fill the universe. Mend it. Illuminate it. Open, warm, feed it, as it does for us. So a garden, a pie, a child's triumph at learning a new thing. Small and without footnote, repeated in household after household.

Brisk wind, warm sun. Summer easing out, not ready to leave.

Tomatoes now, cabbage, corn. Climbing beans and cukes. Chard, beets, carrots. Basil ripe and ready for making pesto. And a steady supply of zucchini. We shall not want.

Heading to rest, put my heart down for the night, mind open and silent to the push and pull of the day. Leaves yellow, and the heat of summer goes on. We should be so lucky as to live in such a time, seasons shifting and the work of ages to be done.

Humid and hot. Summer blazing as it retreats. Closing in on September and all the returns and change it brings.

THE GREEN DRAGON
WRITING THE BOREAL (SECTION TWO)

Spring Boreal Trip, Our Second Visit, Fisher River Cree Nation

Our spring foray is delayed a bit due to flooding, but we get there in high spring, all life at full tilt. Again we rent a van, pile in our bags and laptops, our sketchbooks and notepads (the paper kind). Ken packs in his sound equipment, Mandy her vast array of cameras. Janine has brought us healthful and alkalizing snacks for the road, for our hikes. No end of nuts, sunflower seeds, raisins. We head up the west side of Lake Winnipeg this time, to the Fisher River Cree Nation, just adjacent to the Peguis First Nation, about an hour and a half north of Winnipeg. It has been six months since our winter trip into the forest east of the lake, that cold week just before winter solstice. We are ready to see the forest as it opens itself to green life again.

 This is the spring of water, after last winter's long cold and heavy snow. Everything is wet and green, a little chilly. We had hoped to go in April or May, but all the rivers overflowed their banks, including the Fisher River that runs through the reserve and empties into Lake Winnipeg. So we waited until June, ready to catch the coattails of spring in the forest. Sid drives the first leg of the journey up, trading off with Ken at the town of Fisher Branch, a dogleg turn toward the reserve.

 We ride in a little late in the afternoon; it has been raining, raining. Roll up at last to the motel, and waste no time settling ourselves in, room by room. It's not a long journey, but we are tired nonetheless, and hungry. We head out to Loretta Lynn's restaurant for fresh pickerel from the lake and Loretta's butter tarts, for which there are no adequate superlatives.

I pull out my laptop, start writing.

June 2

Bush and gulls outside my window. Mornings I take coffee outside and walk to the road that separates the house, the motel, from the forest. Is it speaking? A standoff. I stand and look, it stands. I don't enter. Know enough not to enter. But I come anyway to the road and look and listen.

No dreams to tell. But surely we dream here? Does the dreaming forest eat them? Do our dreams go out into the forest at night; are they someone else's now? Do we not know how to dream here? Are we growing new minds at night? Bird minds, poplar minds, mouse minds, frog minds, bear minds? Do the forest spirits eat them? Do we parlay with the spirits at night in our dreams? Are the spirits kind?

The road between me and the forest is a border, a liminality. What do I hear there? Something going on. A non-mind language, or communication. I don't have the words. There may not be any. What does it teach? Does it say no, or ask what is your name? Or, what right do you have to know me? Do you have permission? What is the magic word?

Too busy now, everything is waking and growing, feeding and fighting. Excitation. Exaltation.

The forest does not plead. Nor does it promise kindness. Nor not. It is reaching always for its climaxes, the moment and the years behind and ahead. To be home. Each spring it rises up.

June 3

We walk the lakeshore single file, not singing and then singing. Cleaned bones of a raven, eaten by the rez dogs or a lynx. Driftwood staffs present themselves, smoothed willow branches. Tangles of roots and fallen trees. Sand that sucks things down. Out to the point, brave city souls. What are we looking for? The liminal? How land and water merge? How the forest is eaten by the lake; how the forest defies and lays its branches into it?

But this is spring, we are looking and listening. Water everywhere, and it goes on raining, little bogs in all directions. Mosses impossibly green, full of light.

The lake takes my glasses, sent a leaning willow to push them off into the beach sand, at the juncture of earth, water and air. I am taking lessons in the holiness of the forest. It gives you music, it gives you breath. What does it feed? Does it have a summer spirit? Does it have a spring spirit?

The forest listeners. Us.

The land women: Hazel, Arlene. They are tied to it, extensions of it. They move in it at home.

Elder David Murdock tells us there is no harvesting of herbs in the spring. Harvest in summer and fall. When you go to harvest, it is your intent that is paramount. The plants need to know why you are harvesting them. Tell them it is to heal. He teaches us about the importance of names. There is a name the universe knows you as. The grass and trees know you by your medicine name. What is your name? No one gives you a name. You just have it. Someone, an elder or medicine person, can hear it for you. But not give it.

He tells us the importance of not holding on to people when they want to leave life.

He gives us each an eagle feather.

I have noticed that Elders make themselves naked. They tell you their lives, they tell you the holes they have dug themselves out of in their lives.

David tells us about the importance of relying on friends, the importance of help. You can't do it alone.

I ask myself, is the forest naked like that?

Yet it encloses stories. You stand at the edge asking entrance, offer tobacco. You hear the voices of birds telling their stories. Nothing else. Or the story of how wind moves or doesn't move through it. Beyond, the lake. On the other side. You might hear it on a stormy day. You might not.

It might be the trees in their ecstasy. They might dance it for you. The greens, one laid on another, dark green, lime green. Their patter and swish. Too subtle a dance for the city mind.

Enclosing

That the forest encloses medicine. That it encloses air. That it carries these. That it encloses lives. How would we find the medicine? That you speak to the plants, identify yourself by your medicine name. To the life around the plants you want to harvest.

Name

Dare you enter without your name? Dare I enter without my name? Ravens quarrel in the trees. Call and answer. Warn of dogs, the lynx, cougar quiet on her fast paws. The bear on his last round. Evening, the trees warn do not enter. Spring is full of danger. The rains come fast. Again, ask the forest for entrance.

Moose and bannock in the rain

We are eating moose and bannock in the rain with school kids, teenagers too cool to wear the yellow slickers, but proud of their work in carving the moose, cooking it, making a fire in the rain. A road over a swamped creek. Cow tracks. Birds calling on either side, from the reeds, the shores and shallows. Sid asks, I wonder if moose is kosher. Meat is tender, tucked into bannock. Fire smokes from the drizzle.

Elsie

Elder Elsie challenges us to write music about the slaughter of bears. She shows us her sweat lodge. An owl has been visiting her all day. He is perching on a small tree near her property. She is wondering what he wants. It is a small owl, young. He sits on the treetop chirping, swaying. The day is volatile with rain and not rain, pelting rain. Ken and Sid are drawn to the owl. They stand and watch him until he flies away. Perhaps he has told them what he wanted to say.

Elsie shows us the wards on her property, her trees, coloured cloths tied around them. Elsie is angry, defending herself.

Walking to the forest

Walking up the road to the forest. Over and over. Eagle wheels overhead, turns and heads back into the forest. Days are full of water. Children, teenagers on the land.

Water, the disappearing frog. Ken records the evening songs of the frogs in spring, calling for the dance. The dancer.

Sense of the gift of the owl and the eagle.

Walk to the forest, face the forest, speak and listen.

I stand at the door of the forest. I ask permission to enter. Morning after morning I ask permission to enter. Dare I enter without knowing my name?

The River Pelicans

We come by the small river, visitors. Pelicans fish alongside people, group on a sandbar. Further up the river there are shallows, reedy water. The pelicans glide in, low to the river. We are excited, awed to see them. Not the people living here, they are co-fishers with them. The pelicans aren't afraid, don't spook.

Water

Everywhere. The lake. The river. Creeks. Spring flood, the winter melt off, rain, rain, rain. So watering the spring, forest, lake, the fishers out for spring pickerel. The spring fishing. Pelicans on the river, fishing. The fish spawning. The pelicans nesting. The eagles and cormorants fishing.

Birds

Eagles ravens gulls pelicans. Crows. And the small songbirds from the south coming home. They are heard in the forest. Not seen, hidden in the trees. Their voices only calling. Ravens in the evening forest, establishing who rules. Cacophonies. Then silence. Then again. "Quiet," say the ravens, "our babies are trying to sleep."

Mammals

We hear something, unfamiliar cries. Birds or mammals? We see only the rez dogs. No bear, no lynx, no cougar. Not even a fox. Nothing. Not even their bodies crashing through the woods. Too close to the roads. Houses. Are they even there?

Summer Boreal Trip, Our Third Visit, Jack Pine Lodge

July 26

We came up here through the Icelandic corridor – Gimli, Hecla. All names of gods and heroes, and Hecla, the long-road town, houses all looking out to the lake. Remnants of boats and fishing life still on the shores. Now the lake is polluted in places. We cannot excuse this.

We spend the night at Hecla Resort. Incongruous hotel rooms like spas. Forest colours now full deep greens. Mosquitoes galore. Soft air. Soft water. A hot day, I swam in the glittering afternoon lake. But there were warnings on the beach about water quality, don't swallow, the sign said. Eddies of cold currents in the warm water.

We walk to the lighthouse, through the short forest. Moss on birch, iridescent green on white bark. We pick stones off the pebble beach, all round white limestone. The smooth limestone everywhere. And granite. Squadrons of dragonflies at evening, diving and hovering for mosquitoes. And the swallows diving and swooping for them. Bird voices, unidentifiable, singing. Happy to be in the bush.

Late evening, I'm on the patio of the hotel, drinking vodka and tonic, watching the night come down. The lake with its ring of forest. Water sounds, small trickles of the artificial waterfalls in the hotel gardens, the lapping of waves on the beaches. Always looking out to the lake. I watch and watch as the dark clouds roll in with night and the thunder comes.

July 27

Cloudy, breezy. Rain came in the night. Don't recall my dreams. Skeeters want to eat me. Swallows dip and dive, eat them. Nice economy.

And we drive up to the ferry dock in the wind. No ferry today, perhaps all week. The weather is too rough. So we come down to Jack Pine Resort and find ourselves accommodated and unaccountably happy. Weather sucks: cold, windy and wet.

The bush is next to our hotel/motel. The sheltered bay is surrounded by dark green islands of summer forest. Three small planes in the water at the dock. Clouds moving west to east, a procession. Sun cloud, sun cloud. If you walk into the forest the wind quiets. Shelter. On the dock wind, wind. This is summer, and fire now is the wind and rain. Nothing burns. High summer here. Wind makes poplars spill their stories, chatter all at once. Are they telling the same tale? Do they gossip?

July 28

We have moved to a small cabin for a couple of days, further into the woods.

Up early, don't recall dreams, but the lake and forest are beside us, around us. The lake is a player in this part of the forest. Lungs, breathing, and water feeding, filtering. Limestone cleaning. And granite of the Shield always around you.

The forest remembers your name. You don't.

You walk the roads, and by them there are purple prairie clover and fields of strawberries, raspberries and the thin yellow clovers of the plains. Wind and sun take the bugs away, and you are walking free on the hard roads. The forest floor is soft, spongy, crackles a little underfoot from dropped twigs. You might be walking on moss, on rot. Dead trees are ringed up and down with fungi reaching in to break the cellulose down, and who but the fungi can do this? Send their rhizomes deep in, make fungus babies, ring up, little shells on the stump.

We are berry picking again, a feast for breakfast tomorrow. Porridge and berry mush. Ken off with his bug hat and his microphone. Wanting to pick up the songs of the exquisite little birds we cannot see. Mandy always with her cameras, scouring the forest floor.

Saw nuthatches on the trunks of trees out by the front of the cabin, their tiny quick movements like the gestures of wrens.

We give back to the forest with our scraps. Thank you, here is some food for you too. Not that it needs our leavings, but they are degradable, edible. Someone, something can eat them.

So for these few days we are forest people. Trees surround us, the lake in front. A birch hangs over the veranda. We forage for berries. Ken is the fire keeper. Janine, the house mother, the organizer. Our house in the forest. Mandy documents us, the house, the forest, her cameras open and snapping.

I am just the old lady here, and a helper. Wood fetcher when there is no wood.

Janine asks what did I want to be when I was a child, what did I want to become, and I said I had once wanted to be a pilot, a healer, a teacher, a priest, a nun. And a writer. Yes, but Janine and I didn't speak of that, it was a given. The little airstrip here reminded me of that desire to fly. How do you fly now?

On the forest floor, the little berries, plants snaking through the underbrush, berry bumps like tiny diamonds on the fruit. Leaves, twigs, mosses, tiny plants, tiny insects. You don't know what you might step on. But it is always soft, bearing you up as you walk. Renews itself from its own detritus, moss bodies make new moss, tree bodies give life to the shell-like fungus, the dropped leaves feed the next tree.

Forest breathes in the carbon dioxide, sucks it in, breathes back oxygen, lungfuls. We know the drill, the cycle. Self-organizing, self-sustaining, self-protecting. Forest invites bears, wolves, coyotes, lynx to keep out the unwanted. It is not romantic. But it romances itself in spring. It has only a few months to be alive. Half the year it's under snow, the other half, under leaf.

It's chilly at the end of this July, cool, rainy, windy, little sun, not high summer. No fire, the summer is too cold and wet. Next year, the cycle can wait a year.

Here we eat like health nuts, food all fresh, pure. Too good. But every nutrient must count, be used. So we eat like the forest eats. Waste nothing, no calorie. No junk. No waste.

And we dance with the forest, dance with the mosquitoes. Dance with the heart of it. Find the hum, the beat. Does the forest answer? Listen? Laugh? Ignore you? Does it know your name?

You can ask nicely, you can command. You can bargain with them, make friends with the skeeters. The evening comes and they gather for food again, blood of a mammal, blood of a bird. Air too still, and they dance in the stillness. Dance in the daytime, dance at night. For life. No swallows here to dive for them. Few dragonflies. Who eats them? Frogs? Do fish eat their larvae?

We are preoccupied with these little things. Their sting, their bite, the blood they draw. They will not be ignored. Small things that rule the summer forest. Like ticks in late spring. Skeeter bit me on the ear, drew blood. Listen. Hear. And the drum sings, beat rolls. Builds up the hum. Summer, high summer. Fierce summer.

Longing for a thunderstorm. Noise and light. What happens when the clouds explode, when the air ignites. Fire meets water. Fire meets air, fire meets earth. Thunderstorm: all elements in play. Earth grounds the fire and receives the water. Flash of a dream. Summer calls the elements together in this. Explosion of the clouds. Striking the earth. Soaking it.

Teeny tiny ticks.

I sit outside with wine, watch the lake and the dogs. Eagle comes by and wheels into the forest.

July 29

Mandy and Janine battled skeeters all night; it was the night of the bug warriors. This morning we laughed about it. The ninja warriors killing all night long, trophies on the wall, splatted mosquitoes.

Brief visit from the sun. Then mist again. A fine mist of rain on the lake, can't see the other side in it. Mandy out shooting in the bush, wearing her bug hat. What will this cantata become? Music, pictures, a log of a journey, four journeys. An observation of one year, a full round of seasons.

We ask ourselves what we are doing, we don't know, just listening. Letting the forest speak to some part of us, and us making art from it. Is it me, or is it the forest that calls for these words, those photos, recording that sound. Making that observation. Learning that lesson.

Janine notes the theme of death for her, and healing, learning. We talked of shrouds and green burial. For me, it is all about the name. Who are you? Can you remember?

I can't ask the others what their journey is. But we are in a work together. Work we may not altogether understand. Journey, travel, destination. Here we are our own spirit leaders, spirit teachers, and the forest leads and teaches us. Speaks and sings to us. Elements, circumstance. And me here wanting to drum, but put off by the bugs and the rain. Drum with words.

Ken off with his recorders to tape the rain on leaves. Yes. Mandy knitting. I trusted this, the insight, that we would learn all this together, on the fly, as we went along. How to create together, how to listen, how your ear and voice blend with, speak with, the heart of the forest. How to hear its sacred songs, read the great moving text of its seasons. Too enormous for a few words, but we shall try.

It says, I was here before you, I shall be here after you. Forest comes after glaciers. Peopling and peopling the land. Rock. Nexus, centre, all elements and the axis of spirit, all directions, all seasons. Making life.

Mandy is knitting by the fire. Ken is recording the sound of rain on leaves. Patter. The quiet wind.

First entry for me into this forest was at eleven years old. Just as the body begins to move into puberty. Alive with it. The forest then was adventure, a clear lake, sun on the dock, eating the berries like a little bear, crashing through. Scents of that forest like this one. What calls you at that age? The green, the soft moss, the movement of crayfish in the shallows at the dock. Let the adults worry about bears. We are foragers at that age, like them. Fishermen, cooking on a rusty barrel outside. And no fear then. Living teaches it. So here I am again. Calling back something I left there.

Everything tastes better in the forest, says Janine. The ninja who defeated Mosquito.

On the forest trail: we saw Eagle fly over, and Raven followed us back from the hike. The hike into the forest was short. We sat and listened for a time. To my right there was a portal, a doorway in the trees. Out of the doorway came Deer, with head and antlers and body like a fawn with the markings, but adult. It stayed there, and at one point turned into the head of a black moose. Then Deer again, stag. Visiting us, or telling us something. We sat still for some time. Skeeters swarmed us, hands, feet. We sat still, and the doorway stayed there for some time. There are always doorways it seems, or a few, or where you do not expect them. Then Ken says, look up, and Eagle circles us and flies on. Raven just above us. Calling to another raven a distance away. Ken records Raven. A small bird sings its way into the mix.

Each one of us has a journey. For me, the authentic. The wild. I feel insanely happy here.

Sitting on the cabin deck, I am surrounded by small brown twittering birds of different species. So much flitting, small birds. Small birds.

Trees sprinkling down the leftovers. Dreaminess of the horizon. Mist again on the lake and you know it's rain. Spruce puts out soft light cones. Wish I had brought a sketchbook. How does the forest heal? Sucks your blood. Pulls out the poison like leeches do. Thins it. The skeeters do the work, the little ticks.

We take a walk to the airstrip to watch a tiny silver plane take off. Pelican lands on the breakwater. Sailboat tacking in the distance. And the trees still talking, taking in the wind.

Wind in forest again, evening wind.

July 30

We take to choosing animal spirit cards together and reading them. Pulled Turkey and Lynx today. A gift, secrets, old knowledge.

Janine pulls Hummingbird – Joy. Mandy gets Salmon. Wisdom. Ken pulls Lynx. Secrets, something to be known.

A brief gift from the forest: a glimpse of the sun rising. Then cloud again. No wind yet today. A small spider on the table, visiting. Mandy out shooting pictures, Ken off on a hike. Both bug-hatted. I wanted more light and sun, but each glimpse is a gift. Dappling the forest. A bit of time to drum. Sound of the small planes taking off and landing. Some on land, some on the water.

Time is a gift, the "no obligations" factor here at Jack Pine gives us all downtime and open space. Yesterday, Eagle circled us in the forest, a salute. Today, I hear Raven again, further off in the forest. Calling and calling. Crows hung out in the trees this morning, their beautiful shapes silhouetted against the lake.

In the meantime, here we are. There is a small bird chirping above the cabin. We have to leave it in an hour, go back to the motel for our last night here. And trusting we can travel home tomorrow. We are not where we set out to be but we are where we need to be at this point in time. We have loved this time in the cabin. We leave our blood here in the forest, thank you, and our DNA goes into it. Ticks, skeeters, whatever bites and takes blood. We become part of it.

We move back to the motel for one more night. We have had a couple of lovely days together in this cabin, right on the water. Small brown forest birds. Can't find their names in my mind.

Connecting again is a bit odd. Emails, Facebook, what is going on in the world. What will the forest tell us, what images, sounds, words? High summer this year is all water. No fires. Swift rains and as swiftly, hot sun, then cold rain again. Air moves very fast. Clouds are on a mission. No notice of the world below, creatures like us.

July 31

Raining lightly. We are going home today. Feeling sorry to leave the bush. It has been generous to us, and so have the people. It was a haven when our plans went south. With the wind.

Last night we made a fire on the breakwater. Told ghost stories, Ken played the jaw harp. Leaving music. Shaman music. He was speaking to the spirits, or learning to.

EATING OTHER BODIES
SOME ETHICAL CONSIDERATIONS

Every time I read ethical justifications for vegetarianism, I get discombobulated. Not because I think it is wrong to be a vegetarian, nor for that matter to be one because one empathizes with, or feels for, other animals. Generally prey or food animals only, though, not the nasty predatory kind – one has to wonder what sort of ethics would arise from empathizing with lions and wolverines and sharks and orcas and eagles.

And no, I don't think there is anything wrong with basing an ethical position on feeling, or empathy, and, certainly, a good deal of the ethical basis for non–meat eating is feeling. Surely empathy generates such ethics as feminisms, post-colonial ethics, environmentalisms and so on. It is a good beginning place for ethics. And no, feeling is not the only place from which to start ethics. Though it is *feeling* for food animals which generates a great deal of ethical vegetarianism.

Vegetarians, vegans especially, tend to represent themselves as being, shall we say, hyper-ethical, placing themselves on a higher ethical plane than omnivores. I have even seen arguments that the vegetarian is "further along the evolutionary path" than the omnivore, as though this were a true or even desirable trajectory of an ethical universe. And let us be very clear, by no means are all vegetarians ethical (I believe Hitler was a vegetarian), by no means are all ethicists vegetarians, and by no means are all ethics about food.

Let us say, as a hypothetical situation, that a vegetarian is invited to dinner at an omnivore's house. Let us call him, as he might like to call himself, the "Ethicist." It could be that the hosts of this dinner might be concerned enough about the Ethicist's feelings as to make a non-meat dish; he, however, might desire them not to be so anthropocentric in their empathy, catering to his feelings rather than the feelings of animals they themselves might be eating. But that is their moral choice, not his. If our Ethicist feels like he is going to dinner with the Görings, then he perhaps should decline the invitation in the first place.

Of course, any guest, such as our Ethicist, should let their host know if they do not eat meat, just as one should let one's host know if one has food allergies, or religious dietary rules. As someone who cooks for large groups frequently, I *want* to know these things, and it is not a hardship to have something on the table for vegetarians, or to make a dish without a certain allergen in it. I often cook kosher dishes for my Jewish friends.

Let us say that our Ethicist feels constrained to proselytize his vegetarian ethics at the table. Now, one should *not* proselytize one's moral position at a dinner party, though one should answer truthfully if asked about it. One could then have an interesting discussion instead of a nasty altercation. And no, one should *not* impose one's taste on other people, who are perfectly capable moral agents and can decide those things for themselves. No, one most emphatically should *not* express one's disapproval of the host's choice, no more than the host, or other guests for that matter, should express disapproval of the choice of vegetarianism that our Ethicist is making himself.

In defence and justification of a non-meat diet, I frequently hear the "live and let live" argument. I would suggest that no vegetarian is in a position, nor has a right, to "let" other beings do anything, including eat or not eat meat. People decide these things for themselves. One hopes that our hypothetical Ethicist is not trying to stop other animals from being carnivores, nor putting any kind of moral stricture on them. If we humans are just another animal, then why should one make such a moral distinction? The human animal is omnivorous, like bears and pigs, and always has been. Some animals are strictly carnivorous, yet others are herbivores. Each has an ethical place, and a job to do, in the balance and flourishing of nature. None is more or less ethical, or more or less evolved, than any other. And yes, all predatory animals cause suffering to their prey. Will our Ethicist try to put them all on a vegetarian diet to save their souls, and prevent the pain?

I notice that "live and let live" extends pretty much only to animals that humans eat. Nothing much is said here about mosquitoes or bacteria, for instance. Does the Ethicist refrain from swatting skeeters? Flies? Does he refrain from taking antibiotics if he is ill from a bacterial infection? And nothing at all is said about plants and fungi. Since they are all living creatures, should we not also apply "live and let live" to them? It never ceases to baffle me that ethical vegetarians draw a sort of invisible line between plants and animals, and put them into a hierarchy of aliveness. I don't understand it. Things are either alive or they are not. If it is wrong to kill animals, then is it not wrong to kill plants also? What about fungi, whose nature hangs somewhere between plant and animal? And what exactly *are* the ethical issues around the engineering and mass production of plants? Or, when it comes to plants, are we only concerned with what happens to *us* when we eat them?

Let us turn instead to the question of people who may be accused of not being concerned about the non-human animals' ethical standing. Does this mean *all* animals or just animals that humans eat? And by "ethical standing" do we mean their ethical standing in relation to the rest of nature, or just how they stand according to, or in relation to, humans? How do predatory animals stand ethically in our Ethicist's books? Parasitic animals (and plants)? Will our Ethicist judge them also for the pain and suffering they cause? Or are they "different" or "lacking" something, perhaps empathy, and humans are not really just another animal? One must take a position here, and extend empathy, feeling, to the plight of predatory animals as well – tigers and certain eagles are nearly extinct, as are wild African hunting dogs, wolves are persecuted, orcas are made into toys for human entertainment.

Surely, the *real* moral issue in eating meat is not that it is wrong for one animal, humans in this case, to eat another. You would need to condemn pretty much all the natural world for that, and thereby doom everything to extinction. I suggest that the real moral issue here is the manner in which most of our meat is raised and slaughtered. But I do not find that to be an argument against eating meat, I find it an argument against current methods of meat production. These are two different ethical issues.

I have long believed that we have allowed profoundly unethical methods of livestock and poultry raising and slaughtering to develop around us. Most of us have no idea how it is all done. We have gotten too far away from our food production to have any sense of connection to it, to what it is, how it grows, how it is harvested and so on. If half the energy and indignation that goes into ethical vegetarianism went into lobbying for more ethical livestock and poultry raising, we wouldn't be having these arguments. I would like to point out, though, that similarly unkind methods of production go into raising our fruits and vegetables and grains, and I don't hear anyone squawking about the unethical treatment of plants.

Everything we eat has to be mutilated, robbed, killed. Not just cows and pigs and chickens. Everything, right down to the lettuce and tomatoes and onions and mushrooms in a salad. Mother Nature is indeed red in tooth and claw. That is her great compassion, her kindness. There is food for all. If anyone has a problem with that, they may wish to take the matter up with her.

Now let us say that our Ethicist is in a dilemma as to whether he should try and influence the behaviour of others. In answer, let me tell you a story. Some years ago my mother attended an international poetry conference in Malaysia, and, along with some other Western writers, was asked to dinner at the home of a respected Malaysian poet. This man was a Muslim and had two wives. Naturally my mother did not agree with this practice but, of course, had the grace not to criticize her host. Other Western guests at this event were not so gracious, and loudly denounced their host for having two wives, at his own table, and while cheerfully eating his food. My mother was appalled. It was not that she disagreed with their position regarding polygyny, but, rather, that they so rudely tried to *impose* their ethical beliefs on someone else, *and in his own home*. The two wives were of course completely mortified, as was the poet himself, but they were far too polite to throw these obnoxious jerks out, who were, after all, guests. How would our Ethicist respond to this type of situation? Was it a moral evasion for my mother to keep her opinions about polygyny to herself? There are always going to be situations where one has to weigh one ethical obligation against another.

So I am asking myself, upon what does our Ethicist really base his vegetarian ethics? Feeling, yes. I do not disagree with him that empathy should be extended to beings other than ourselves, beings which are under our power and in our control. But given that they are, and I include here plants and fungi as well as animals, how should we treat them ethically? That any creature, whether wild or domesticated, is liable to experience pain and suffering in their life is a given. That includes ourselves. We live in a culture that refuses to accept pain of any kind, and we throw as many analgesic drugs at it as we can. Pain, suffering and death are somehow seen as affronts. So I worry that ethical vegetarianism relies too heavily on a sentimental, anti-pain bias. I will qualify here that there is unnecessary, gratuitously inflicted pain, as well as normal pain and suffering that any embodied life brings as a matter of course. For our Ethicist, however, animals must not suffer or die, at least not while we are watching.

But look around. Nature is not sentimental. Shall we criticize nature for not having vegetarian sensibilities? Are we humans a cut above nature, ethically speaking? Just a wee bit superior? Set apart from the rest of it? More ethically evolved? To imagine it is morally superior not to eat anything animal is to fly in the face of natural ethics, and to pass moral judgment on every predatory animal on the planet, and some plants. This shows no respect whatsoever for other life forms, nor for the natural order.

And being selective about which creatures we empathize with, such as animals over plants, is, I suggest, a kind of moral evasion, and it is one of my main quarrels with ethical vegetarianism. Let's carry this argument to its fullness, and apply the idea of "let live" to *all* living things. Obviously, if we are to "let live" in that extreme sense, then *we* don't live, because we can't eat anything. If, on the other hand, we are just animals, then like every other animal, we must depend on the lives and deaths of other creatures – animal, fungus and plant – to continue the existence of our own species.

For good or ill, we have entered into contracts with certain other animals – dogs and cats, for instance. These animals have a contract with humans to help us with our lives, and vice versa. And since we became agriculturists and herders, we have made contracts with certain plants – certain grains, fruit trees, root and leafy vegetables, etc. – and animals, such as cattle, pigs, goats, sheep and certain birds – chickens, turkeys, geese, ducks, etc. – to be our food. It seems to me that it is the contract with these animals and plants that is being breached. Not because we are eating them, but because we are not treating them kindly and with respect as they grow, and in the manner in which we kill them. We are not giving the animals good lives, freedom to move around, proper food, sexual enjoyment and companionship. And in this day of GMOs, I don't even want to consider how appallingly we are breaching our contract with plants.

Most of us don't have to kill our own food, unless we are farmers or gardeners. I am a gardener. Every summer I force genetically engineered seeds to grow unnaturally, and abundantly, in neat little rows. I thin, or kill, them if they are growing too close together. Then, just when they are in full ecstatic growth and really enjoying life, I come along with a big sharp knife and a basket, and take their lives and the lives of their children, day after day, for my table and for my freezer. I hack off their leaves, I twist off their fruit, I yank them up by their roots and I chop off their stems. They can't run, they can't scream, at least not so I can hear them. They can't protest, and say NO NO! I want to live! Yet they are living beings like the rest of us. How can I morally justify this murder, this pain and suffering, this theft of the next generation? I do what I can. I ask permission of the plants for their bodies and their lives, and then I thank them for their sacrifice.

I have often wondered why there is no plant rights movement. We only seem to care about what looks like us. Respect or care for other living beings appears to stop at the end of the animal kingdom. I say this is a "feel good" position. I say that if we respect life, we need to respect all of it, and not create a hierarchy of aliveness in order to make ourselves ethically comfy.

All life feeds on other life, one kind or another. But it would be ridiculous, not to mention cruel, to require all predatory animals to become herbivores, because it is considered more ethical. Meat eating does not need to be justified. Meat is an ethical food source, along with plants and fungi. We are not ethically above or more evolved than other animals. We live with them on this planet, we take positions of predation and husbandry, like many other beings do, in the food chain. If we choose to ethically separate ourselves out, set ourselves above, other animals, then we create a false otherness for ourselves. We may judge non-human life as either less than humans and lacking a moral compass, or as not participating in evolution, thereby having only an instrumental role in the human moral universe. And by reserving our empathy for food animals, we create another moral hierarchy in the animal kingdom, and the false role of victim for prey animals.

There is no moral nor ethical ruling in the fabric of the universe that says humans should only live where there is vegetable life because it is "gooder." And no, we don't know better than those who choose to live, say, in the North where there is little vegetation but abundant animal life. People lived with and in and of this environment in a balanced ethical order long before tofu hit the North American markets.

Let's be clear: we kill other beings every time we pull up a carrot or chop up a mushroom. There is absolutely no reason to stop eating other animals. It is not a path of moral evolution. Where we have failed as a species is with respect to how we raise and treat the animals we eat. What has happened, I think, is that there has been a moral, spiritual devolution *in us,* in how we raise our plants and animals for food, how we harvest them. Our ancestors lacked nothing in terms of food-gathering ethics; if anything, they had a better understanding than we do, and much better ethics in this regard. And if we stop understanding the relationship between eater and eaten, and I fear we have, then we fail as a species. It is not *that* we eat or milk other animals, it is *how* we do it, in what spirit we do it. That is the ethical failure. Our ancestors understood something we do not: that food gathering, whether hunting animals or gathering plant matter, is a matter of a respectful partnership and exchange. We have utterly forgotten how to recognize and value the generosity of other beings – plant, fungus or animal – in sharing their bodies with us, so we can live. We are very lucky indeed they have agreed to these partnerships.

Fall

SEPTEMBER

Fall is always the shift to open light, and the deep calling in of green life before frost.

Berries in the thatch, birds eating little grapes and shitting purple. Wrens in the garden, the hawthorn, a hummer checks the salvia and fills up. So much bird life in the garden, trying to fly into windows. North windows in the fall.

Rough wind, cool night. The dark comes early now, slow walk to the autumnal equinox, and then the turn toward winter. But the roses still bloom, and the vegetable garden is giving its great yield of tomatoes, zucchinis, beans, beets. The fall lettuce crop has been planted.

Pale rainbow on grey clouds. Sun in the West, going down. Safe in the rain. It's the sound of it on the roof, pattering past the windows. Safe, safe in the rain.

Forecast says rain again today. Cool and wet, green as you please. Geese vee-ing, honking, following the river south. A few jays squawking in the trees. Squirrels scolding, rushing about, squabbling. A few little songbirds still stopping to catch a garden meal on their migration. And here's us trying to keep up with the harvest.

Big elms are yellowing, more light coming through the dense leaves. Music all day, the sound of air going amber. There is morning sun after the night's long rain. Clear blue sky. Calling what is in the air to attention, sweet scent of plant decay, wet earth, voices of the birds flying south, sunlight on trees. Prayer of the rabbit, spell of the raven.

Near frost this morning, the tender things in the garden must come in now. We are close to the equinox, a week or so away from the balance of dark and light, the turn toward the year's evening, another growing season completing itself.

Wind is bowing the rose bush outside my sunroom window. Too much remembering. September calls up sorrow. Just the dying of the season, or is one entering a wind that has a sad job to do?

Yesterday, began to take up the garden, and so much vegetation the earth makes in one season, in one small garden! Bush beans are done, broccolis finished and cauliflowers struggling to curd this odd wet year. Tomatoes shrivel their leaves as the fruit ripens. So goes another season, slowly putting itself away.

Sun rising later and later as is right for this time of year. Warm days for us this late September, leaves falling fast in the autumn winds. Turning the spirit now inward, readying the house of the mind for the long dark nights to come. Want to walk and walk in this air. Still sunny, a little breeze. Crisp as only fall air is. Crunch of leaves underfoot. Walk with the sun, with the fading sun to its rest.

Rain pulling down yellow leaves, makes running sounds like marbles in the gutters. This garden lush now, even at the equinox, and we are safe inside. Contemplating change as the season turns, the moon opens all her light, the sun balances and then shifts south.

Walked the neighbourhood just before dark. Homes closed against evening and the coming cold. Last blaze of colour in the gardens, fall flowers and late roses. Patter of drying leaves about to drop, and the trees pulling in their sap. A vee or two of geese looking for water or a field to dine and sleep on. Dog walkers on the roads, political signs on lawns. Lights already on in windows.

Robins at last mobbing the rowan tree, munching down the berries. Winging back and forth in little squadrons.

Sun bright today, garden damp from frost turning to dew. Desire rising to be out in it. Autumn such a siren, air filled, overflowing with the goodbyes of the plants, they're calling you to bring them out of the earth, take their gifts in.

Thinking of the slow walk of the Green Man to death for the winter here. Plant by plant, leaf by leaf, fruit by fruit, as the sun walks south on the horizon.

OCTOBER

Another warm, sunny day, we are promised. When autumn is beautiful here, it is nonpareil. Sun slants lower now, lights under branches, floods through the south windows. Air is sweet with falling leaves, gold, crimson, chocolate.

Recalling the sacred stories of the season, autumn, the dying of the green, the cull of animals, winter coming. How generous the gods are, the earth, how loving.

Crazy warm days, October in like a lover, and me inside at the keyboard, wanting hands in soil, feet bare. Rivers are rolling too high and fast, like spring. Seeds flying in the breezes, birds still lingering in the trees.

Thinking again of the forest. Walk up the stones, air biting a bit, from time to time a raven croaking, leaves just turning and the forest quiet. Bears out there somewhere, it's rut season. Elder takes out his drum, sings welcome, calls to the directions. Our eyes closed, spirits already in the wind. Rough land making itself out of stone.

Sun already up, forecast twenty-six degrees Celsius. Just outside the sunroom window, a tiny bird on a leafless branch. Robin chases it off for the vantage point, makes a beeline for the rowan to score the best berries, first pick of the morning. Already, there is a breakfast crowd jostling in that golden tree, knocking berries to the earth, bouncing the boughs. This kind exchange: food for the birds, seeds scattered for the tree.

Sunday again, circle of the week come round. So quick, time passing, even in this long warmth. Tomorrow is Thanksgiving, and we are in short sleeves still, and still slowly harvesting. A gift, this warm autumn.

Going up to twenty-five plus today. And it's October. Nights still cool or cold, but days glorious. Dizzy with ecstatic sunlight, air's fullness. Want to stay out all day, walking it, playing in the garden.

Touch of frost again. No more garden rescues, just bringing it all in now, putting the plant bodies to the compost. Earth is a kind eater of what is left when things die.

Warm for fall, a little rain early morning. They say sunshine for the day, and I will gather it in, windows open, arms out to it. Sun riding south, its arc shorter, lower in the sky, a little more each day. Slow farewell, peace, yearning.

Such silence in the sky, the birds gone or going. Maybe a last late vee of geese hugging the river in their flight, maybe a crow calling from the roof, maybe a sparrow. And oh, heard Raven, his forest voice speaking in the city trees. We are so close, so close.

Still raining this morning, though they promise sun later. Good day for writing, and a long autumn walk. How much light there is in trees as their leaves turn, how warm and open. Like walking in bowers of light, walking under them.

The furnace chugging now, cooler nights. Perennials pulled out and potted, herbs that don't winter: sage, rosemary, parsley, chives. Geraniums. Ivies, spider plants, oregano, fuchsias. In the sunroom now I'm surrounded by living green, shelved, hanging from the ceiling. To remember the past summer, to look forward to the next one. To listen to their spirits whisper all winter long.

Tonight full moon. I am the Blue Hag, the Winter Queen, by moonlight walking into the land again. I bring you the long darkness, the sparkle and ache of frost, deep rest of winter, the turning inward, horizon yearning toward light.

Hag leads us down to the dark, the resting earth. We walk into the sleep of trees, deep rumble of their winter dreaming, their singing in the earth. Listen.

NOVEMBER

Always waiting in November. For snow, something to change, for the holidays to come. Or the universe to shift in some way, currents to turn in the auras of planets or stars. Sun to touch the horizon and climb again. Constellations to find their way up the sky and speak, urgently, about the workings of celestial time.

Says it will snow today, and it might, might not. Frost on the grass, a scurry of squirrels and small birds. Boeuf en daube in the oven, potatoes steaming for mash. Comfort that winter brings, only winter.

Thinking now toward Yule, just over a month away, and the snow is promising to come soon. Sun folding into the south, stag moving to catch it in his antlers, hold it for release at the solstice.

Sun and wind today, still warm for this time of year. Making parsley root and leek soup, from the late harvest, all roots. Keeping us grounded in the earth as the winds shift the currents of the world.

Winter doing a slow saunter this year, no storms, not yet, just cool days, cold nights, a dusting of snow. Grey curds of cloud. Sun lower each day in its southern arc. So we turn toward the solstice to come, this month of waiting.

November moon balancing, moving toward full. Orion now a giant in the south sky of an evening. Blue jewels of the Pleiades catch your heart of a clear dark night.

First snow yesterday, thick, white curds on the grass; thick, white sleeves of it on branches. It will melt quickly, gone in a day, but oh.

Full moon. Putting the land to rest tonight for a winter of sleep, with thanks. All insects gone to ground, or south. Few birds still here make a winter living from berries, the spruce cones, wild grapes, seeds scattered in the light snow.

Snow curds on the grass, on roofs, trees. Sparrows in the eaves, waiting for dawn. Sing open the doors of winter, you.

Sun pale this time of year, declining south in its arc, a bit more each day. Leaning toward the stop, the turn, of earth to come. How ancient the sacredness of this gesture to people of the north.

Snow walking, your boots sing as you go.

Today we begin the Yule season with music, choir singing it in, and us open to the energy of their voices, harmonies, rising up of sound, the light of it dancing in our bodies.

Here comes the kindness of snow, protector of earth in winter. Light as feathers in air, thick on trees, roofs, fences, on fields, in forests. Calls up primordial love, memory of ancient hearts, arms of the gods around us.

Winter is in. Now we count down to the solstice. Small sparrow on a bare branch, hunched in its winter feathers.

Last day of November. Tomorrow we start to decorate for Yule. Remembering solstice in the forest. Stars talking fire in the deep nights. Evergreens and naked birch trees talking under the snow. How cold the air, and how bright the sun on the white river.

THE GREEN DRAGON
WRITING THE BOREAL (SECTION THREE)

Fall Boreal Trip, Our Last One, Wallace Lake, the Petroforms

This is our last trip to the forest, autumn the last season to taste, smell, hear before we begin to set words to music, choose the pictures of our journeys for the online gallery, before we weave the sounds of the land into the fabric of the *Cantus*. We began this journey in winter, when forest life was immanent, waiting, resting, struggling to stay alive; we moved, visit by visit, through the opening and ripening of spring and summer months. Now we are ready to walk with the forest as it folds in to sleep again. We come without prejudice, without preconceptions; we come, really, with ears and eyes open, ready to be struck with wonder, knowledge, understanding, to be carried into its music. After this, we will set to work in our city homes, minds reaching out, time after time, to the boreal.

October 3

Like coming home, the dusty gravel road winding, twisting to the mining town of Bissett, the town at the end of the road, about ninety minutes northeast of the city. The forest more beautiful, more exotic each mile, more homelike. A dusting of yellows and buffs on deciduous trees, the conifers deep green. Late fall this year, it should be gone, the foliage. Floor of the forest light with the yellow and red leaves. Janine grew up in this town set in the forest and on the ancient Precambrian Shield, and it is in her blood, her mind, her imagination.

Our first foray of the trip is to the Bissett cemetery, a short trek into the bush, just west of the town. It is a small clearing in the forest, a few headstones sprinkled around in the cut grass. A tall evergreen, its branches dripping with grey-green lichen that hangs like Spanish moss, presides over the home of the Bissett dead. Under it, a long bench faces the family graves of the Wynnes. We came here last winter to find Janine's mother's grave, but it was deep under the snow. Today we find it without trouble, its cement veneer now cracking from weather and age. Janine has not seen it in thirty years.

The cemetery is quiet, save for the rustle of the drying poplar leaves, which only stops in winter. Everything is now frayed and yellow at the edges, red in places, small bushes with mostly eaten berries. Carla and Mandy are photographing a bird kill on the road – something has killed, defeathered and eaten a grouse. I pick up a feather and pocket it. There is a wing, now all bones and a few tip feathers. Whoever has eaten it has cleaned the bones well. Carla and Mandy take snaps of the scatter of feathers, here, there along the little trail in to the cemetery. Ken is hunting sound along the edges of the bush.

We look at each gravestone. Some Janine knows, and she tells us a story about the deceased. Some are older than her memory of Bissett. Only one is older than Bissett itself, a small grave of two children who died sometime in the 1930s. Long ago in town memory.

After dinner we take a walk to the lake, just as the sun is disappearing into the west, a little south now, we are past the equinox and moving toward the dark time of the year. Janine wants to show us a rock near the edge of the lake, one she is familiar with, a place of both refuge and discovery for her when she was young and listening. The rock is gigantic, and scored deeply by the glacier that passed here some fifteen thousand or so years ago.

It's cold, too, and, with the chill air, I begin to regret not bringing my parka instead of all these layers. Still, I am warm enough for the night.

The town is built on rocks, granite broken and scarred, striated by the glaciers. If you climb high enough you can look over Rice Lake, its far shore almost touchable, and now spattered with patches of yellow over the deep green of the evergreens. The small islands, the still autumn lake. Little float planes tied up at the dock.

I wonder if the stone will keep me awake at night with its hardness, the banging inside it, its density speaking and speaking, stories of its life, the lives of the trees, the lichens, the animals that have passed here through their lives, long and short, the creatures who have crawled on them, lived in their cracks and crevices, the human lives that have come and gone around them. Now the people live in houses perched on the Shield, the immense shield of the Canadian boreal forest.

The soil is young, Carla tells us, only a few thousand years old, very thin, the forest grows in the cracks, the fungi and lichens break it down, granite as dense as any stone in the world. These organisms can break it down, digest it, slowly. This is the long cycle of the forest, the growing on rock, breaking it into soil, individual trees, whole stands come and go in their tens and hundreds of years. The forest cycles are seventy or eighty years of youth, maturity, old age and burn. And the cycles of the years, the seasons. Cycles within cycles. Lives within lives.

As this autumn slowly arrives, I think of the spring forest with its excitation of nesting and birthing, the bursting of new leaves, the homecoming of birds from the south. I think of the slow summer of insects and fruit, the rains and the deep greens of climax. There is a calm now as the birds gather and feed to leave on their southward migration again, all the young of the year matured, ready for the gruelling flight south. The winter birds, the year-round birds, still cluster and quarrel at feeders, in the trees, nabbing the ripe berries that still cling to bushes, picking at seeds close to the ground.

We know what it will be like in winter, we have been here. Quiet and full of light from snow. Short bursts of intense white light. Days shorter than nights.

Tomorrow we are going to Wallace Lake. Tony and Carla went there for a field trip as undergrads, and are glad to return these decades later. We want to see the new forest, the growth after the burn cycle. It has been about twenty years since that forest burned, and we want to see how a new forest grows, what communities it makes, what grows first. Is it the birch bringing its light and short life, then the slow spruces, pines, tamaracks?

Carla asks how these visits to the forest have changed our view of it. I can't say it has changed my view, I answer, but the visits have opened and expanded memory. We lived in the northern forest for two months the summer I was eleven turning twelve. I recall the scents, the height of the trees, the heat of clearings, the sound of feet on moss and cracking twigs. Picking berries, listening to the lap of lake against shore stones.

Watching, hearing the little float planes take off and land on the lake. The scent of the summer forest. The air full of oxygen. Small dusty roads. The taste of bush food: fresh fish caught off the pier, cooked outdoors in a bit of butter and flour; wild raspberries pulled off the bush and popped into the mouth. Bursting with juice and tasting like rubies. Distant wolf voices. The always-there frisson of knowing there are bears somewhere around, or moose in their craziness charging out of nowhere. Rain in the trees, dripping long after the rain has stopped. Those strange black edges and holes in leaves, bacteria already eating them as they grow. Spindly forest floor plants reaching for what sun they can catch. Treetops tossing with ravens, crows, eagles. Their calls fixing in your heart and memory, your body, somewhere primeval.

Driving in through the twisting roads, lined with trees, safety and home.

October 4

Today it's Wallace Lake, due south of Bissett, where the forest burned crazily some twenty-five years ago. There had been a research station there for boreal study, an extension from the University of Manitoba, before hectares of it disappeared in full flame. The fire took the research station with it. Now, it is a forest regrowing, renewing itself, following its own cycles, answering its own needs.

After Wallace Lake, it is catch as catch can. Discovery, impromptu stops, pulling off the road and piling out wherever the land calls us.

I brought my drum, to drum with the forest again. Ken his jaw harps. Perhaps we will make some music. Sid will return to Winnipeg this evening with Tony and Carla.

But the poetry of the forest will speak. As we enter the renewed forest of Wallace Lake, we learn the names of the plants from Tony and Carla. Their stages of life.

Carla had spoken yesterday about poison ivy, how dangerous it is in all seasons, even winter, yet the birds can eat its berries. White columns of berries on low stalks. We had walked for a bit by the dam at Powerview-Pine Falls, into the trail leading south, before turning north and east for Bissett. Leaves of small shrubs, crimson with pomegranate-coloured berries in columns, clusters at the top of stalks. The deep shiny scarlet of rose hips. Scant leaves on the plants. A lone coot still moving through the reeds behind the dam. Oak trees, young ones, no acorns on them. The road east and north beside us.

Shield country bangs with the energy of the stones, the granite. Grey, pink, covered with lichen and striations, gouges, like trails of animals digging the dirt. The forest is creating soil, homes, food, lives, oxygen. We breathe the forest. It lives like its denizens along rivers, by lakes. Water. Shorebirds, water birds. Small songbirds. The great predators and scavengers. The eaters of plenty of insects in the summer. These are now almost gone, though there are a few left, stragglers in the lateness of the fall. Still food, still warm enough for them to feed before flying.

Geological time is short here, everything young in those terms. The rocks may be old, but the landscape is creating itself now. All the creatures and plants together, co-creating it. Bears still roaming and eating, rutting. Soon they will find dens and bury themselves in sleep for the winter. Fat, and the females pregnant. Still a few berries for them, though not many left. Always a few garbage bears about. But we are not looking for that. Nor are they looking for us.

Guardians of the forest. Bears, wolves, ravens, spirits. In winter, just ravens and wolves. Bears gone to bed, eagles gone south. In the fall, the eagles are still here, they will marshal soon, I am told.

At the lake, we pile out of our vehicles, head out for a morning jaunt to the new forest. Much of this forest is young: Jack pine and poplars, birches. Carla says the Jack pine is a healer after a burn. Its seeds are hard cased, and need the burn to come open and sprout. We arrive at the campground, and locate a trail. Here the forest is deep in Shield country, and as we climb into the forest, we climb the oldest mountains in the world. Pink and grey granite, streaked in places with white quartz. The forest grows on it, out of it, unmaking it into soil.

We walk through the fall flora, heads down, the forest floor is where the action is. Mosses, lichens, fungi. We find a multitude of late mushrooms. Carla explains them to us. Mosses, Tony knows, and there are mosses we have never seen. Lichens that surprise us – a little red one called British soldier for its red coat, among the grey and green. The lichens make a living off the rock, break it down for the tiny Jack pines growing in the cracks of the old granite mountains. The soil is thin, and the trees spread their roots laterally to hold themselves in place, to find nourishment in the sparse layer of earth.

We finally reach the shore. Someone has made a firepit. We lie back in the brief sun and Ken silences us to listen to the fall forest. A distant float plane. A squirrel chattering. A crow somewhere over the trees. The lake barely lapping. Silence. Voices in the distance, someone cracking twigs. We wish we had brought our lunch with us, but we break silence as a few drops of rain, mist almost, begin to fall. We hike back, hungry now. The shoreline of this lake is still dark green, only a little yellow in it.

The campground has picnic tables, firepits. We haul out a smorg of breads, spreads, cheeses, corn chips, cookies, apples. Carla and Ken make the fire, Janine and Mandy lay out the lunch. It's chilly, cloudy. Fall weather. We eat standing up around the fire. After lunch, Carla demonstrates how to balance a stick on your hand. We play the boreal stick game, everyone having a go, balancing the stick. We are laughing at our own silliness.

We decide we will find some hidden falls on the road back, but first we stop at a granite quarry, just past a small airstrip. Here we feel like we are climbing the ancient mountains again, and pause on the highest rock to view the new forest. The vista is stunning, greens, yellows of fall.

On our way out, we find another bird kill. Our second this trip. Yesterday it was the remains of a grouse, feathers and bones. Today we find what seems to be a raptor. We don't know the species, so I take some feathers to identify with my bird book when I get home. I take the feathers also for the bird. You do not know what gifts the forest will give you, sticks, stones, feathers. Today I took a stone shaped like, and sharp as, a knife, a granite shard from the quarry. Perhaps it was dynamited off.

We took photos of ourselves on the rocks, surrounded by the forest. Striated rocks shot through with quartz. There is such laughter this trip. Sid wants to know how to identify the trees. Carla and Tony teach him about the Jack pine and the birch. Sid is nimble as a mountain goat going up and down the rocks, over the stumps and fallen stems of the trees.

What is noticeable is the silence of the forest in fall. Few birds left to sing in the treetops. Some whisky jacks amuse us at the waterfall. So much else is ravens and crows, no mating to be done, no territories to defend. We end at Birch Falls, or at least that's what Janine thinks it is called.

A hidden river in old forest. Black winding river, the tall thin trees of old forest lining the banks. We scoot down a steep bank to the first station of this, what becomes a kind of pilgrimage. Old man's beard lichen dripping from trees like Spanish moss. We move further down: here, black sharp rocks make a chute, the black water begins to funnel, plunge through them. The water becomes rapids, leaping over unseen dark stones, down to another chute, a fall, foam rusty with tannin. A pooling, then down more rapids, then a short fall into a cauldron of black water.

You know this is a holy spot at the first stop, the winding river in the protection of old trees, some now yellowing with the fall. You know you are in a holy place at the second station, first falls, the white foam on black water. You know you are in a holy place at the third station, the rapids and the black cauldron. The water slips away then almost unseen down a small bed, and under an arbour of branches that almost hides it. Another world beyond.

Secret and you hardly notice, the river goes its own way past this doorway. Black water and white foam. The old trees watching.

Hardly a bird in the trees, no mammals. You don't even hear them, though there must be bears in this forest. Weasels, raccoons. Mice. Like winter, but the colours, the colours. No season for colour like this one.

A few water birds, still hanging on, a flock of coots in the reeds of Rice Lake, a small bird darting into the empty reeds, leaving green into the buff of winter. A few geese. Beaver crossing the unrippled inlet home.

If there is one audible voice that spans the seasons of the forest, it is the voice of the raven. Even in the fall, the winter, you will hear him, hear her. Sentinel in the quiet seasons, in the loud ones.

We walk again through the town of Bissett; it is decaying houses and memory, empty houses, doors coming off hinges. Who are the forest people who lived here? Who still does? Miners, opportunists. Like the cemetery, the town is almost empty. Open lots where houses were, where there was a store, a movie theatre. The old mountains still under them, still making soil with the trees. They have to blast through this granite as though it were mountains, and that is because it is, was. Laid on its side, this old range, the whole Cambrian, geological time so long we cannot imagine it. Or we can, or at least put numbers to it. But the slow movement of rock, air, water, fire. Earth making earth. Creatures of air, wind, water; the rapids; the rain; glaciers; the lakes that come and go. The fire of cleansing and renewal, forest cycles of burn and grow. We saw the new forests today, along with the old.

What voices has the forest? Animal, bird, water, wind in trees, cracking of twigs under foot, the fall of trees, the roar of fire. The patter of rain. The slow, slow groan of rock as it ages, moves, sings in its long geological time. We hardly hear it, but listen, the old mountains call to the trees; the fungi; the lichens; the creatures that walk on them with hooves; with padded, clawed feet; who scratch themselves and wear them down over time.

We had thought to connect, make music, drum and harp, but we fall asleep too early for that, take time alone in our rooms. There is no drumming in a motel, there are other guests.

Tomorrow we go to the petroforms, look on these ancient sacred sites. What are the animal forms for? Clans? Calling the spirits? Remembering? These old mountains were alive and singing in their prime, there was no bear, no otter, no creatures of the woods as we know them. They had not yet evolved. They had not yet come into being. Vague forms in the lower world, that is all. Or were they already there, waiting to materialize when the time became right? When the old creatures had run their course, died out. And now, where are they? Hanging in museums, only bones.

What will come after these?

The forest breathes, makes air for the air breathers. Makes soil for plants that feed animals that feed plants.

October 5

The sacred story of soil making. The cycles. The characters, intrepid plants, fungi, beings who pioneer and prepare the way, make it possible. The lichens, the fungi. Each being is a community of beings. Tree with its fungi, its lichens. Co-operatives. Some die along the way. Fungi the decayer, breaking things down, making things possible. We eat the carrion makers. Plant carrion. Calling the food into the tree, breaking it down when it dies. In the fall you consider the closing, breaking down of things. Bacteria in leaves, fungi in the roots, praise them indeed.

We make our way to Nutimik Lodge in Whiteshell Provincial Park, all forest and lakes. The Whiteshell as it is commonly known. We settle into comfortable cottages, each facing the creek, each with a firepit and picnic table. Two per cabin. Our first family meal of the trip, we cook, we sit down together, we do dishes. Tonight is Ukrainian food night, perogies and cabbage rolls, and raw veggies. Some from the garden. We have arranged to meet Ron Bell who will take us to the Bannock Point petroforms this evening, at seven. It will be dark. Ron is the keeper of the petroforms, and is an expert on them, knows where many are that have not been seen.

He comes for us right at seven p.m., and we take my drum with us as we leave. We enter from the highway; it is only a short walk up and into the site. We stop at an entry rock where offerings of tobacco and other gifts cover the rock. We each put our pinch of tobacco on, asking permission to enter. Ron walks us counter-clockwise in the dusk; we can make out some of the forms, turtles, snakes, pathways on the rock. He wants to take us finally to the circle, where there is a medicine wheel, and a grove of trees hung with cloths. He tells us again the story of how Turtle became the back of North America, Turtle Island. And he allows as how as he himself is Turtle Clan.

There is a stone circle around the grove, and we enter it at his invitation. Ron tells me to sit down on a flat rock and play my drum, and I do so. Ken asks permission to play with me on his jaw harp. We play together until we are done. Ron wants the ancestors to come and dance, but the cloths don't move much. Yet we see their lights in the grove, in the trees around. Small blue lights winking. They are here, but we are strangers. I would have danced with the drum, but Ron had told me to sit. Not my culture, and I am here at the sufferance of those who know it. Whose world and structure it is. But I would have danced had it been mine.

By the time we are done, it is dark, and he takes us out the way we came, careful to say thanks with a pinch of tobacco. We agree to meet the next morning to trek out to the Tie Creek petroforms. He tells us we may dream or have visions from this visit. I sit out before bed for a smoke, watching a beaver wend its way to the docks on the opposite shore.

October 6

I don't recall any dreams from the petroform visit, but I wake up worried that I won't be able to walk the trek into Tie Creek. Worried my knee will give out. Ron has told us it is three and a half kilometres into the site by foot, then the same out, and we will be walking the site itself.

It is overcast and cool when we arrive at the pathway in. Ron walks with his eagle wing ahead, announcing us. The path is as wide as an ATV, and is double rutted. The route is littered with puddles and little bogs, but green still, and bounded by forest on either side. Through the bush we can see Precambrian rocks rising in the forest, still a little bit mountain.

We manage the walk with wet feet, and very muddy shoes. It is not as difficult as I had imagined for my bum knee. It takes us an hour and a half to make the three and a half miles, not kilometres. Suddenly we come out into a clearing, and it is all rock. We are at the entryway of the petroform site. Big boulders are placed perhaps as sentinels, or as welcomers. A buffalo stone. A drumming stone. Ron drums it to see if we can enter the site. Luckily, it drums out the affirmative. He tells us once he took a group of people in, and when he drummed the stone for permission, it was not granted. He trekked the group back out again without them having seen the petroforms they had walked the three and a half miles to see.

Permission granted, we leave our offering of tobacco in the hollow of the stone, then walk in. The site is a rolling sea of boulders punctuated by standing rocks, tree clumps, lichens and mosses.

We walk further, and enter the portal of two stones, into the land of the ancestors. We are walking west. The petroform site itself is now bounded by a huge chain-link fence ten kilometres around. This became necessary to prevent people from running over the stones in ATVs and snowmobiles, but Ron has the key.

How ancient it feels here. Some estimates for the age of the forms, or for some of them, is about ten thousand years. After the last ice age. If that is the case, then these forms were laid down before there was a boreal forest. Given the amount of time it takes to generate soil enough for the trees to get purchase and food, the forest is young in comparison to these forms. I think about Carla telling us, showing us, how the soil is created, slowly, bit by bit, by lichens eating the rock, mosses, fungi, the litter of vegetation. On the site, and throughout the forest, tipped trees show us exactly how thin the soil is, even after ten thousand years, the roots growing outward, not down. Soil holding these trees cannot be more than four to six inches deep. The soil is made on, and partly out of, the old mountains of the Precambrian. The forest grows on it, and breaks it down.

Like the Bannock Point site, this one also has coloured cloths hanging everywhere from the trees. Ron blows his eagle whistle to let the ancestors and other spirits know we are here. We have been walking a long time, so we pick a site out of the wind and build a small fire, sit and eat a backpack lunch. The rock is blackening with new soil, and with age. Perhaps pollution. While we rest, Ron pulls out his drum and sings the welcome song. I am happy that I can sing along with him. He goes to the nearest and biggest form, and speaks to the directions, calling with his drum.

It is like being in a slightly different dimension. The place is heavy with the age of itself, its energies upon energies making dense the air, and we are all knocked a bit off world. We walk them all, and we see that the stones are now overtaken by lichen, forest beginning to claim it. What can you do? I think of the knowledge stored in the stones, and in the huge worn-down mountains we are standing, walking, sitting on. It is all stone. A sea of stone.

Ken wanders off to take some sound, and Ron amuses us by making us sing songs. Ken gives up; we are too noisy. Ron fills and lights his pipe and we smoke to the ancient spirits here. They are everywhere with their ancient languages, their old knowledge of stars, land, rivers, lakes, ceremonies, magic. Teachings of how to live well. Animal energies. Clan totems. A gathering centre of tribe upon tribe. Here there is still knowledge of how to speak with the ancestors, how to dance with them. Call them to your ceremony, your dreams.

There is a petroform that Ron tells us is a thunderbird, with the outline of a sweat lodge inside it. This one, he says, is the most potent.

Before we leave, Ron gives us a tied red and white cloth to hang in the trees for our project. We put our hands on it, put our hopes and wishes into it. Janine and Ken hang it in a pine tree, high.

The rocks of each petroform are not large, but are arranged in patterns. Some are animal shapes, some are oblongs with straight lines, some are piled on each other. Stones all mottled with grey-green lichens.

As we entered the site the sun came out, and while the wind moves constantly on the big rocks, they reflected back its heat, and we looked into the blue sky as we walked the range of the site.

This is fall, quiet in the forest, quiet at the resort, and very quiet here at the site. A few whisky jacks squealing. The ever-present sentinel, the Raven, calling from time to time. Patter of the drying leaves of the birch, the poplar. Yellow everywhere in the green. We walked in on a highway of stone. We walk back out on the same highway. It has been walked for so many centuries it is almost as though it had been built that way, but you know that the ancients found it, recognized it, and walked it the way it led them.

As we walk back through the forest to our parked van, the light of the afternoon sun shafts through the trees, pools along the path. It is easier to see into the forest with this much light. Ron spots some wild ginger and stops to harvest it. We take a rest on the still soft and moist mosses along the road. I can't remember the names of all the plants. He has been instructing us along the way which lichens are good for which illnesses, how a certain bracket fungus can be lit and inhaled to cure a headache.

By the time we arrive at the van, my knees are aching, I just want to sit down. But we are deliriously happy, and we do not know yet how the ancients may have spoken to us, or not, and we don't know what dreams may come after.

We make a fire in our firepit after dinner, chat a little. My body tells me to lie down. But not before I see beaver swimming in for the night from somewhere down the creek.

October 7

I am stiff this morning from the walk to the petroforms. The walk in was long and delicious; the walk through the forest to them; the portals in, the highway of stone to somewhere very sacred.

It turns you inside out. A way in to speak to the ancestors, to learn the stars. The forest is encroaching. Why now? Something has happened, as it should. The way in, the walk, the portals, the drumming stone. May we come? The knowledge must come out.

Today we are leaving the Whiteshell, the forest. We have done our year of seasons in the boreal. The last trip of this phase. We make a daylight visit to the Bannock Point site, by ourselves, and walk it this time clockwise. First to the big circle with the medicine wheel, then around to the animal forms, making a circle back. Snakes, frogs, turtles, no end of turtles here. Mandy snaps some pictures in daylight, permission granted for this by Ron. We leave the site and head down through the Whiteshell, out of the forest for home.

Along the way we see more eagles than we had seen all three previous trips. Wheeling over a goose sanctuary, two of them. Then about ten scattered from a roadkill as we drove by. Then another couple from another roadkill. I had seen one from our Nutimik cabin while watching the creek. It flew over, carrying a fish in its talons. Forest life. Geese came home at dusk, the beaver too. Silence of the autumn all around us. The sweet scent of leaf decay. Sun further south as it rose and fell. I had bought a bottle of good Cabernet Sauvignon when we got there, knowing I could not finish it. The rest I gave to the land, the forest, in four directions, as thanks for the year, the four trips out, the learning and listening, the pleasure and danger of the boreal.

The choral piece, Cantus Borealis, *premiered with the Manitoba Chamber Orchestra in April 2011. It was a beginning, an opening of the voice of the forest to the ears of the city.*

It takes us a year, more or less, after our four visits to the forest, of writing, meeting, testing, editing, testing again, scrapping, editing, rewriting to bring the *Cantus Borealis* into being. And in the end, we hear it: the drums, the wind, the water, the voices of the birds, the path of fire and the footsteps, the panting of wolves. In our shaping of this piece, we walk again, season by season, through the trees, by the lakes and rivers; picking the berries, swatting the bugs, clambering the rocks, watching tiny planes take off in the middle of the bush. Watching the sun disappear into a lake. Standing under a cold black sky, wide-eyed at the enormous stars. Bundled against a bitter wind, walking anyway into the wild.

RETELLING THE STORY OF NATURE
HOW TO RESTORE AN ETHOS

The natural world is collapsing around us, folding under the weight of humanity's demands. At this writing, Lake Winnipeg has been declared the world's most threatened lake, overfed with phosphorous from sewage, agriculture; overused by Hydro projects. At this writing, whole mountainsides of conifers are dead in British Columbia from an unprecedented infestation of the pine beetle, from the monoculture planting of trees for the logging industry. At this writing, salmon spawning grounds on Canada's west coast are disappearing, the forest streams they call home destroyed, a direct result of clear-cut logging. At this writing, there is a battle in the Amazon between the government and the indigenous people over damming the river basin for power. Forests in Malaysia are being razed by lumber companies, destroying the habitat of countless creatures. One could cite example after example of a global breakdown between humanity and the rest of nature.

This breakdown is made easy by the narratives, or stories, we create and tell ourselves about nature, as we forget to listen to, to read, the narratives of the natural world. We have forgotten how to be taught by the land, the oceans. Instead, we have invented narratives that position humans as apart from nature, or superior to nature. We have created narratives of nature as instrumental to humans; of nature as fallen, as made "just for us"; nature as mechanical, etc. In our time, we work and live from within scientific, commercial, political and religious narratives, which, for the most part, have in common their careful stripping away of nature's subjectivity, of its natural generosity, its natural state of grace, its status as naturally ethical. These narratives describe nature as an object, and reserve subjectivity (soul, perhaps?), ethical superiority and privilege for humans.

I am calling these narratives, which are now naturalized into our psyches and our world views, false narratives. These false narratives are cemented in our laws. They inform our expectations, our behaviour toward the rest of nature, toward the land. These naturalized false narratives grease the wheels for approaching the natural world with unbridled rapacity and greed.

If such damage can be done through these multiple narratives, then they must be countered by narratives that will bring humanity back into alignment with nature. I propose that we reach back into pre-urban narratives of living with and on the land, and with its beings, and in this way restore the heart of our own lost land ethics. I am proposing nothing less than the restoration of an elder ethos, one that understood and strove to keep balance, which grounded itself in respect, reciprocity, gratitude. The land may hold part of that narrative, and somewhere in our own deep past, or deep mind, our lore, we hold the other part.

Is there a tool, or a method, with which to restore an ancient ethos as practicable in our own time? And since the damage has been, and continues to be done in large part through narrative, can an ethos also be restored through narrative? In my research on ethical restoration, I have yet to come across a method specific to restoring an ethos. This has led me to try and develop a methodological approach for doing so, and for pairing ethos restoration with narrative. To that end, I will be unpacking and using a number of terms toward that end, including "restoration," "narrative," "aboriginal." And I will use a geometric model called the *torus* as an illustrative model for narrative and ethos restoration.

The terms restoration and narrative have separate lives, so, first, I am bringing them together here to create a third term, "restoration narrative."

In the context of this essay, the term restoration narrative will refer to the process of the restoration of an ethos. The first trajectory of this restoration reaches back toward what I will refer to as the aboriginal, an ethos arising from an originary source.

Restoration of ethos simultaneously addresses restoration of narrative, as, in this model, the one carries the other. As an example: from the perspective of neopaganism, the skewing of historical and religious narratives by a series of patriarchal discourses is remedied in part by reaching back in time past the appearance of those discourses, for an earlier, more true (or ethically useful) narrative that restores the lost aboriginal ethos. We can view this as the search and rescue trajectory.

The second trajectory is forward from the past and into the present: recovery and redeployment of the aboriginal ethos, with redeployment as the result of the forward motion of this trajectory.

However these trajectories are not linear but, rather, take a spiral form: the source of the ethos is always already there. A linear trajectory would look like a line of stitching, which starts at an originary point and moves ever forward in a straight line, away from the beginning; a spiral trajectory would look more like sewing on a button, with a back and forth motion as the needle and thread loop behind the button and come back forward again, staying in the originary place but never making the same stitch twice.

I'm going to unpack my terms restoration and narrative separately to uncover and discover their dictionary meanings as separate words. This allows me to choose definitions most pertinent to this inquiry. I can then bring them together into a single term, which can be used to examine my texts. Since restoration is the first word in line, I will begin with it.

Restoration

To "restore" is to "put back." According to the *Britannica World Language Dictionary*, it is "1. To bring into existence or effect again: to *restore* peace. 2. To bring back to a former or original condition, appearance etc.: to *restore* a great painting.... 4. To bring back to health and vigor. 5. To give back (something lost or taken away); return."[1]

The *Canadian Oxford Dictionary* defines "restore" as "1. Bring back to the original or former state by rebuilding, repairing, etc. 2. Bring back to good health etc.; cure. 3. Give back to the original or former owner; make restitution of. 4. Bring back to dignity or right; reinstate. 5. Put back; replace."[2] As "restoration," it is "the return of something to a former or original state (*building restoration; ecological restoration*). 2. The act of returning something to a former owner, place or condition; restitution."[3]

From the *Compact Edition of the English Oxford Dictionary*, Volume II, "restore" is "1. To give back, to make return or restitution of (anything previously taken away or lost.) 2. To make amends for, to compensate, to make good (loss or damage). B. To set right, repair.... 9. To recover, to revive."[4] "Restoration," then, is "1. the action of restoring to a former state or position," and "4. The action or process of restoring something to an unimpaired or perfect condition."[5]

In all cases, to restore is to put something back into the originary state or condition. The definitions fall into two categories: those with obvious ethical content (for instance, restitution, setting right, putting back, reinstating, reviving, recovering, repairing), and those with a gestural or performative, "hands on," meaning (reconstructing, for example). For my purposes, both categories are of interest: the ethical and the gestural.

I am not interested in those definitions that seek to re-present something in its original state or as an original structure. I am not reading restoration as reconstruction, as replication of form. I am looking, rather, to definitions of restoration as "setting right," "recovery," "revival"; we are, after all, considering the restoration of an ethos, not a building, say, or a landscape, or the trappings of a bygone culture.

For purposes of this discussion, I understand the term restoration as ethically positive, and I expect the theme of healing to unfold in terms of both restoration and narrative.

Narrative

A narrative is a telling, a story. Its verb, "narrate," is "1. to relate, recount, give an account of" and "2. To give an account, make a relation."[6] There is also "1. *Tr.* Give a continuous story or account of"[7] and "A narrative is a telling, an accounting, a relating, a continuous story. A narrative can be written or spoken."[8] The terms "relate" and "relation" support both narrative and the ethos that is to be restored in our inquiry texts.

"Relate" means "1. *Tr.* Narrate or recount," "3. *Tr.* Bring into relation (with one another); establish a connection between.... [Latin *referre relat-* bring back: see REFER]."[9] To relate a story is to bring back, restore, make connections. A narrative is in that sense already a restoration. To relate a story is to narrate it. A narrative is a bringing back; a connecting; a bringing into relationship of things, one to another; a *restor(y)ing*.

A restoration narrative will mean a story, a telling, a relating of continuous events whose themes seek to set right, revive, bring back to right and health and vigour something lost, stolen or taken away: a restor(y)ing. A restoration narrative can therefore be described as a healing story. What our [hypothetical] narrative will seek to "bring back" is a truthful accounting/relating of a truthful aboriginal ethos.

Aboriginal(ity)

Since I stipulated that the trajectory of our narrative is toward restoration, then I need a term to describe the end result of that restoration: from what, and toward what is this restoration being directed? Contemporary paganism, for example, has been described as a return of an ancient wisdom or golden age, and could be characterized as return to, and restoration of, an originary ethos that its narratives seek to bring back. Since I am calling these narratives restorative, and the restorative trajectory is both toward and out of an originary condition, I am introducing here the term aboriginal as descriptive of both the source of the trajectories and the object of restoration. In order to locate the end point of this narrative and to identify the object of restoration, I am choosing to restore and recycle the term "aboriginal."

Currently the term aboriginal has come to refer exclusively to flora and fauna, including humans and human cultures, which have lived a long time in one place, and which precede detectable colonizing settlers and invaders, in particular those whose arrival and effect can be documented in historical times. I want to clarify immediately that the term aboriginal does not in this discussion refer to specific points in time, such as prehistory, nor to specific cultures in time, such as contemporary indigenous peoples, nor to ethnic identities. Instead I am returning to its Latin origins to unpack, reclaim, restore, recycle it.

In Latin, *ab* means "from" and *origine* means "origins," beginning. So, aboriginal is "from the origin." An aboriginal is indigenous to a location. Since, at first glance, the trajectories of restoration are both backward and forward (this is how it looked in the beginning, this is what we want to restore), then restoration is *ab origine*, from, and of, the origins, the source (location), and back towards it. However, the location of origine in this narrative will not be found at a particular point on a linear map of time, but rather, at/as a centre.

Britannica World Language Dictionary describes "origin" as "1. The commencement of the existence of anything, 2. A primary source. Cause."[10] Therefore, a narrative theme of creation can be included as a locus of origins.

The impetus for restoration in our narratives is that something has gone wrong, there is alienation from the ethos, from origine, or that the ethos has been broken. In the pagan narrative, the intrusion of patriarchy pushes aside and supplants a right order (which kept the universe in balance, environmentally and socially), resulting in social, racial and species oppression, and environmental disaster.

So for purposes of this discussion, the terms aboriginality, ethos, origine and source will refer to each other, though they are not synonymous.

Narrative Shape

We need a narrative shape that will accommodate both apparent linearity in a story (beginning, middle, end), and the respective arcs of both the larger narrative structure and the embedded action narratives, which propel the story. At the same time, we need this model to describe restoration as theme and story arc, and as ethical trajectory, not just as narrative (story) or as restoration, but as restor(y)ing. Considering these requirements, I am choosing the torus as narrative and restorational model.

What is a torus? "Rotating a circle about a line tangent to it creates a torus, which is similar to a donut shape where the center exactly touches all the 'rotated circles.'"[11]

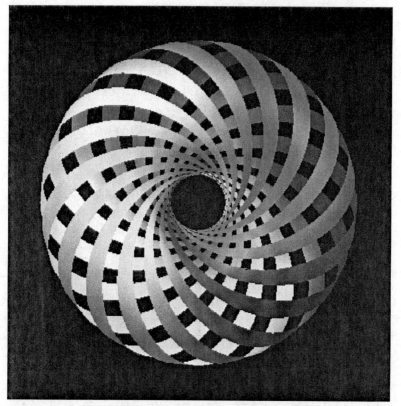

Image courtesy of Thomas Banchoff, Brown University, Providence, RI

The shape of the torus best describes both narrative shape and the shape of restoration in this endeavour. The general configuration of the torus is circular, so it has no appreciable point of beginning or ending. However, it is not flat but multi-dimensional, and its visible body is made up of arcs that move in spiralic fashion to form a donut, which is made up of outward and inwardly moving circles or spirals. The centre appears to be empty, and emptiness also surrounds the shape on all sides, and fills the space not taken up by spirals or rotated circles. The spirals can be described as either moving away from, or returning to the centre, but always at some point touching it. There is always a moving away from and returning to a source, which I am designating as origine. Origine will not be placed at the beginning of restoration but in the centre of it. Likewise, it will stand at the centre of narrative. The ethos can then be described as aboriginal.

Temporally, the torus model includes ordinary space-time and a temporality, or perhaps is beyond temporality but is informing and affecting it, such as an electromagnetic field does, or an event horizon, the non-place-time where the energy events occur in the energy of that field. This model allows for the situating of restoration and aboriginality outside a progressive model, outside of modernist hierarchies of progress. It will liberate the term aboriginal from its social, racial and political meanings, and restore it to its own aboriginality. The narrative, and restorational movement, will therefore not be designated as polycentric, but rather, as polytrajectoral.

The centre of the torus donut will represent the aboriginal (or source) ethical condition from and toward which the multiple trajectories of the narratives move: they spiral rather than move in straight or curved lines. The spiralic lines of the torus will then represent both narrative and restorational trajectories from and toward one (narrative-ethical) centre.

Restorational Ethics

We need a theory of restorational ethics. Because there is no appreciable body of theory on restorational ethics for a non-object such as an ethos, and because environmentalism is concerned with survival, I am looking for inspiration from observations and theories developed by environmentalists. While restoration is fundamental to the environmental ethos, methods and rationales for restoring vary all the way from instrumentalism to humans (value as it serves human needs) to the intrinsic value of nature and natural phenomena themselves (Robert Elliot's *Faking Nature: The Ethics of Environmental Restoration*, for example). Value, however it is constructed or sourced, is informed by, and informs, the ethics of environmental restoration, and vice versa. The effect of a restoration may be the same (saving an ecosystem, let's say) but the values that drive the action may be divergent. Conservation of wild birds and their habitats by Ducks Unlimited may be driven by the needs and self-interest of hunters, but nevertheless, wild birds and their habitats are conserved.

Restoring the value of survival to equal that of salvation (in the best scenario, replacing it) would restore the ethos of aboriginality. In terms of biological restoration of, say, a wetland, the impetus for restoration comes from humans. Why are they intent on restoring the wetlands? What value are they supporting or fulfilling? There is always a practical and/or self-interested reason, such as, in the above example, more ducks for hunting. There might also be an aesthetic reason, it looks nicer, say. Or a biological reason, such as the importance of diversity. Diversity of wildlife is not of interest to say a factory farmer, and the restoration of diversity may be antithetical to the farmer's interests. In whose interest is it, and why is it being done? What theory is the restorationist using? Are they working from an ethics? If so, where does that ethics arise from? Where does the desire to restore come from?

While Bruce Rawles, Arne Næss and others have argued for basing environmental ethics on knowing by intuition, other writers, including Mathew Humphrey,[12] argue that intuition is unreliable, and not a suitable basis for the development of an ethics, in this case, an environmental ethics. However, Humphrey's dismissal of intuition notwithstanding, I would argue with Næss, Rawles, etc., that all knowledge is at least in part based on intuition or something like it. Their application of intuition to starting environmental ethics may or may not be arguable in terms of value development, but it points boldly toward a fundamental human ability (to intuit, and to know by intuition), and gives it pride of place in terms of locating truth claims about the relation between humans and the rest of the natural world. I don't want to argue here about the reliability of intuition (or not) nor how best to define what intuition is, but rather, to acknowledge it as a first line tool of importance for environmental ethicists and restorationists.

I am particularly interested in Alan Holland and John O'Neill's treatment of conservation/restoration questions as narratives. While we are not seeking to restore objects such as ecosystems, nonetheless, many of the same kinds of ethical questions arise with regard to restoration. Holland and O'Neill pose this question regarding ecosystem site restoration: "To which natural state do we want to restore it? To its Mesolithic state perhaps? Or to its natural state during the glacial, or interglacial period? To which interglacial period, exactly? And why this one rather than that?"[13] Since form changes for a site from era to era, it is the substance of balance and diversity, what they call "ecological health"[14] that must be striven for in a restored ecosystem. Since I have identified restor(y)ing as healing, and restoration as bringing back to health and vigour, looking at ecological restoration in terms of health rather than formal reproduction is pertinent to this method of restoration of ethos.

The issue of temporality (time location) in restoration, for example, raised by Holland and O'Neill, speaks to the question of substance or content: what is and what should be included in the restoration. Since an ethos is not an object, it does not have to be locked formally into space-time, it can be restored as the substance of what is actually put back (in their model, called ecological health). The term substance here needs to be fluid enough to take on multiple forms. For example, any given geographical site can suggest a variety of substantial alternatives for its restoration. The question is how to choose the right temporal (time period) and material alternatives (which flora and fauna) to restore to a site in order to effect an aboriginal state in terms of diversity of species, health of the land and so on.

The substances of the ethics, what is included, what is restored, does not depend on form: consider humans returning to a "natural life ethics" by living in caves, and hunting and gathering, for example. Restoration to aboriginality of the ethos (or the ecosystem) is not dependent on such time-specific forms for restoration, but rather, on the content of the ethos or ecosystem itself: for example, the health, balance and diversity of an ecosystem.

In terms of "the problem of how best to continue the narrative [of a given site]," Holland and O'Neill ask, "What would make the most appropriate trajectory from what has gone before?"[15] They state that "conservation is…about preserving the future as a realization of the past… [It] is about negotiating the transition from past to future in such as way as to secure the transfer of maximum significance."[16] The issue of temporality in restoration both is and is not a factor in this definition – the intrinsic value of a site or system is atemporal, but narrative trajectories reach back and forth in time (to the past and future) from the present. I am interested in Holland and O'Neill's concept of "the transfer of maximum significance" in the "realization of the past."[17] Here what is transferred can be said to be atemporal and non-formal; it is not parroting or aping the forms of the past, but rather it is a transfer of significance, or substance. The continuation of the narrative is, or should be, restorative, from past to future through retelling/restor(y)ing it in the present.

In the above model, restorative narrative is a continuation, while in my model, narrative, though certainly a continuation, is also a retelling of the past: it is a healing/restor(y)ing not just of the site in the present but also of the past itself. We can do this by critically rereading the current narratives and then retelling it from a different beginning with new truth claims. The Holland and O'Neill narrative trajectory is a backward and forward arc only, while mine is spiralic. However, the utility of the Holland and O'Neill model is that it inscribes narratability into nature itself – a site can be read as a continuing narrative – and that this narratability can reveal points for ethical consideration. It is in, and out of, the narrative itself that the ethos reveals itself. In this ecological narrative model, as in all environmental ethical models and tools, survival is always the bottom line, informing its tools and its discourses, however divergent they may be.

"Survival": a postscript

In general, this term refers to living past the death of someone, or through a catastrophe. "Remaining alive; living on."[18] More recently survival has been used in conjunction with the mechanisms of evolution (survival of the fittest); and most recently, it has come to refer to the continued existence of the phenomenal world, for example, the survival of a species, or an environment, or a person living through a psychological trauma. In my essay "Ethics and the Goddess,"[19] I considered the question of ethics and survival within the context of an immanentist theology, grounded in the female divine. I looked at the question from the perspective of a Western construction of Goddess. I wanted to consider points of ethics and justice with regard to natural law, or as arising from natural law, and whether, or how, human ethical/law/behaviour coincided with religions/law/ethics as grounded in the divine. The most fundamental biological law of nature is survival, and I believe that the first ethical imperative for all embodied life is to survive. Biological law is an ethical law.

In that essay, I worked from the assumption that ultimately all ethics and ethical practices arise from the necessity of survival, in particular, the various ethical dilemmas highlighted by ecofeminisms, i.e., the connections among devaluation of women, matter, nature and a whole issue of others.

I have not changed my mind on that imperative, that all life seeks to survive. Narratives of conquest, of ungrateful entitlement, of instrumentality, of separation, of disconnection, of human superiority do not support the survival of any species, not even our own, they support only the survival of individuals, or of corporations. If ethics is based in survival then it must apply to all life, not self-selected and self-defined life. We have lost the humility of our most ancient ancestors, who understood this, and saw themselves as part of a circle or wheel of life, not as sitting at the pinnacle. This is what we have forgotten. This is the ethos, and the narrative, which we must restore in order to survive.

Notes

1. *Britannica World Language Dictionary*, s.v. "to restore."
2. *Canadian Oxford Dictionary*, s.v. "restore."
3. Ibid., s.v. "restoration."
4. *Compact Edition of the English Oxford Dictionary*, s.v. "restore."
5. Ibid., s.v. "restoration."
6. Ibid., s.v. "narrate."
7. *Canadian Oxford Dictionary*, s.v. "narrate."
8. Ibid., s.v. "narrative."
9. Ibid., s.v. "relate."
10. *Britannica World Language Dictionary*, s.v. "origin."
11. Bruce Rawles, "Introduction," The Geometry Code, accessed April 11, 2013, http://www.geometrycode.com/sacred-geometry/.
12. Mathew Humphrey, "Intuition, Reason, and Environmental Argument," in *Moral and Political Reasoning in Environmental Practice*, ed. Andrew Light and Avner de-Shalit (Cambridge, MA: Massachusetts Insititue of Technology, 2003), 45–76.
13. Alan Holland and John O'Neill, "Yew Trees, Butterflies, Rotting Boots, and Washing Lines: The Importance of Narrative," in Light and de-Shalit, *Moral and Political Reasoning*, 224.
14. Ibid., 229.
15. Ibid., 221.
16. Ibid., 221.
17. Ibid., 221.
18. *Compact English Oxford Dictionary*, s.v. "survival."
19. Katherine Bitney, "Ethics and the Goddess," in *Life Ethics in World Religions*, ed. Dawne McCance (Atlanta: Scholars Press, 1998), 9–20.

Winter

DECEMBER

In praise of the hawthorn tree in winter. Red berries against white and the grey of graceful branches. Beautiful, the spirit is happy there.

How to sing the forest, how the forest sings itself. Where it sings from. What voice does the Jack pine have, what trees are there? What the air is like, water, sound of twigs underfoot. Creaking of tree limbs. The lake in the forest. The house in the forest.

So near the forest, so near the solstice, not winter coming, but half gone. Crawling down the horizon with the sun, leaning into its southern arc.

Winter solstice: earth stops, tips forward again in her dance with the sun. How utterly ancient is the celebration of this holy event, how universal. How many stories have been told about a child of light being born, a stag, a raven releasing the sun? How many names does it have now, has it had reaching back to the prehistoric past, lore now forgotten? Suffice that for our time, in the lore or science, the great gestures of earth and sun are sacred in themselves.

My holy place is my house, the land extending, snow on the fields, ice on the river. Crows, raven coming and going in the white sky, chickadees singing in the winter trees.

Now watching two little girls sliding on a rink in their boots, grinning ear to ear, bundled in their parkas. A winter delight.

JANUARY

Beautiful snowy evening. At last, at last. Snowed all day, and the neighbours out with their shovels, happy as larks cleaning. It doesn't do, a snowless melting January. Doesn't do at all.

Soft snow again today, wide flakes, coasting in broad arcs downward, circling like gulls when the sky says rain. Ask the Old Mother to shake that pillow again, her featherbed. We should dance in it, turn and turn, lose our minds in that slow white ecstasy. And don't imagine it is not service, to dance with falling snow.

One gets used to living against the wind.

Wind-whipper of a day, treacherous sidewalks from yesterday's melt, now frozen under a thin blanket of snow. How fresh the air now, how packed with oxygen.

Cold night, snow coming. Rabbit tracks in the garden, little hollows where he has made a snow nest for the night. Chickadees huddle in the bare hedge, against windows, under the eaves. Cat prints on the pathway to the compost heap, mice in there keeping warm. Even now, in January, life in the snow garden. Food for the rabbit, the winter birds. Shelter.

Chinook rolling in, across the prairies, into the Red River Valley. A few warm days for skating. A little respite from the bitter cold.

Wind blowing where I live, hauling in from the south, warm air and what they call here "precipitation" – could be snow, could be rain, could be ice pellets. Thinking of a walk in it, same wind that whips the rose canes scratching against sunroom windows. You would think the world had had enough of this shilly-shallying weather. What next, a shift in the rivers? Turning back in their beds? Raking the waterways dry of their creatures? Who will answer for all this?

Walked in it, yes, that south wind. Quiet on the ground. Wind high in the empty trees, sough sough sough. No birds risking their wings in this one.

Hard south wind this early morning, fifty kilometres per hour. Bringing a warm day up with it, bit of a melt as it dies down to daylight. Crazy weather, crazy.

Sunny, warm, a slight melting. January on its head.

FEBRUARY

Wind down, no clouds and stars clear as bells. Clean snow falling all day, still now, having found earth. You pull winter out of your heart, it falls back into your bones.

Snowing. Light, a bit wet. Softening the air, cleaning it, the land. Small boxes of land we live on in cities, little squares and rectangles. A patchwork, so to say, but the birds ignore the boundaries, make each enclosure their own feeding grounds, each building a place to put nests, to hide in the cold.

Cold morning, snow blowing. February heading out like a winter lion. We are a month ahead of ourselves this year.

Snow coming again. Now how kind of the sky, remembering us, our winter drought, the need for ground water in the spring. Not enough this year for a flood, but enough to soak the roots of trees, enough to send up their veins, make leaves. Already buds forming.

THE WATER PROJECT

Pinawa, Manitoba, on the Winnipeg River

Knowing about my involvement with the Boreality project, working art with the forest, an old friend emails to ask if I want to join a new venture he is generating: a gathering of artists of many disciplines to speak of, paint, sing, narrate, dance the voices of water. This is The Water Project. First foray is out to Pinawa, Manitoba, for a weekend on the Winnipeg River, near the Seven Sisters Generating Station, which harnesses that river. Pinawa is about an hour and a half northeast of Winnipeg, and was the site of a research station for Atomic Energy of Canada until 1998.

So here we are – musicians with fiddles and drums, two poets, a playwright, a storyteller, a fabric artist, a ceramicist, a painter – sharing a convention centre with what seems like millions of jovial and shiny-cheeked young über-Christians, here for their annual spiritual pump up. Genial enough. We are a bit befuddled by the energy of their enthusiasm.

February 4

We came here to talk about water. The treasure of it, power. We talk about the damage. We do not have words enough. We talk then about our own land, the forests, lakes, rivers and the oceans. About ancient Egypt, where it was a sin to interfere with the Nile.

Water all around us here, frozen as snow and ice, it is February. It will take fire to break it, the fire of the sun. Fire and ice make water. Water makes life. The world is, or may be, a hyperocean. The whole of the earth is fundamentally an ocean – water, seeking to express itself in all life forms, on land, in the air – the world ocean, repeats itself, makes itself, renews itself, everywhere. Blood. Tree sap. Tears. Spit. Snow. The algae bloom. The dead water, the live, the salt, the sweet, the brackish. Purple water. Green water. What is in our blood that remembers these colours together. The blue, the green, the black waters of a rushing river. Majestic, coming down between trees.

Hyperocean on land: our bodies, all life. Trees. The miraculous wearing down of rock, eating of rock, rivers running through it. But you see, the passion there is in water. Desire and unity. The drive is the ocean in your blood seeking itself. Repeating itself. Not letting go. Drum it and you will see this hyperocean. Fungi, not plant, not animal, but both, the between being. The hyperocean partnering with the fungi. The sea eating rock, old bodies. Sending itself out and eating, drawing it back in.

Algae, fungi. The old life forms bridge land and ocean, water and earth. Always poised at that place. How water comes on land, brings life. How land supports ocean, water, underneath, sand, mountains, hard rock, makes shores and edges of them. No one without the other. At the bottom is the earth, below the ocean. And on the earth, water remaking itself in all life. Or life in water. How do we then understand this? Such a balance. And how does water form, why is it here? And why is it the medium of life?

Lakes are young but they feel older than the sea. Rivers come and go. Change course. Feel young, moving. Small rivers and creeks run in the hearts of forests, deserts. Move water from place to place, fill the oceans back up. Bring fresh to salt, sweet to the salt of the oceans.

I must go to bed soon, but will dream of water, of the molecules of it. A composite chemically, but a single element. Ask the element water. Ask the spirits of the rivers, the lakes. Ask them.

Water can exist in all three forms: solid, liquid, gas. And above, on and below. Ground water, aquifers. Deep in the earth, just below the surface. Underground lakes and rivers. Water on the earth flowing as rivers, streams, heaving as oceans and lakes. Lying still as snow and ice. Above as vapour, as rain falling.

February 5

Started the day with a walk toward the water, the river. Ravens, in their many voices, talking, and the answer of the crow. Chickadees. Birds of morning. Five ravens gather and chat in the sky. Deep croaks.

The sun rose and disappeared again under clouds. Snow now is crusty from early melt. Water in the snow, and we are walking on water.

Talk among us this morning of adulterated, denatured food, of Bannock Point petroforms, of children. And of healing stories. *Just listen. Water speaks as snow and ice.* Ravens everywhere, they are the guardians and the voice of the boreal. Every season, never failing. Patrolling.

Listen. You said ravens, you said chickadees, winter birds. You hear with the ear of the body. Listen too for the trees. We who live here always moving in the night, the day moving in the shadows. Guarding. Watching with the wolf. Peace. Hunger, light coming and going. Water everywhere.

Bears still sleeping here, still winter. Here in the east, it's the black bear. In the west, the grizzly, the spirit bear. In the north, the polar bear. In the south, the sun bear. They are sleeping, yes, but waking up soon.

Children in the hallway of this place, the centre where we are staying. Doors opening and closing. How do you send yourself out to the land? Too much noise. Voices in the hallway. Rest in the forest.

You came here to listen, so listen. So listen. Voices out there. And what do you see? Bend in the river or the lake. Trees: oaks, birches, spruce, and is that willow.

We came just to listen. What does the forest say about water? Holding it now in ice, snow. This place disjointed. Something is pushing against me, or I'm not entering something.

You cannot ask for poems to come gushing forth like spring melt. You cannot ask for that. This home of the heart is sleeping, it is not your toy.

You want to walk again, then, old woman. Find the heart of it again. Want to hear. White, dark green.

You know the stories of the forest mermaids, the limnades, the naiads of lakes. Spirits who save people from drowning. A person I know well, when a boy of about eight, hit his head and fell into lake water too deep for him to swim. He remembers nearly drowning, and then the sudden presence of a being lifting him up to safety on the shore. He opened his eyes to no human. I have heard other such tales from friends, acquaintances. Lore of the forest mermaid.

Give back, give back. Forest has asked what will you do with it? And we are telling it through our voices, the art, story, music. Not enough. No, not enough. Dance it. More.

Forest asking me, *who are you. How do you live? What are you doing here? Justify yourself, what is your name?*

Went to walk into the forest of oaks and birch, pine. Walked to the river, and sat with it. Spoke to the directions, the animals, the air, the birds and trees. Sister of the bear, sister of the raven, sister of the owl, sister of the wolf. *So what did you hear?* Voices. Trees dark across the white lake. Wonders. Urgency. *Give back.* Generosity of the forest, the land, not understood. *Give back. Sit with us and listen.* Lichens on the trunks of trees.

We made a trip, an excursion to the Seven Sisters Generating Station, an old hydro dam. Reservoir quiet, white, vast; water rushing through the station like a natural falls. Snowmobilers taking up the space, back and forth, back and forth. Smelly and noisy. Why do people need those things?

Saw the blue-green of trees on the edge of white, water open and water frozen. Saw bulrushes dried along the bank. Birches loosening their skin. Strong oak, the famous ironwood, dark wood, furrowed bark. Empty arms. Islands in the lake. Open water. Speaking and speaking, heads up to the sun, those trees. Looking south to the sun. Forest says, *give respect, what is it you want?* Home and shelter. Room to live. Breath and breathing. Are you calling me? Yes. *Come home. This is where you return. Bones.* What do I give then. *Bones. Back. Remember balance. No balance in the water we keep making, it is always fouled. Melting north. Ice going. Remaking themselves, all things, water in the blood.* Look at that. Forest and water. All one. The bogs, the rivers and streams. The lakes.

In the forest, the trees hear you, the Old Ones all calling and speaking. *Come. Take out the drum. Play for us. Water in the dam, water in the soil. Water flowing and resting. You can find your way. Yes. All who come here become part of here. Our land, our bodies. Your DNA into us. Ours into you. Food from the land, breath of the air. Why is it fouled?* Speak. Berries, the forest answers. Water and berries, bark. Yearning of the forest. Yearning.

We, the artists and singers, we walk in the same space, speak the same language. Laugh at the same jokes. Create the water project. Water is bigger than this. So we make our art. We are the "Scotia group": singers, poets, painters, musicians, storytellers, playwrights; we all live on or near Scotia Drive in Winnipeg, near the wide coils of the Red River. All of us running ourselves like water. *You think it is enough just to listen but now speak. And write it and draw it. Water. What is it?*

Quilts, pelicans, stories, songs. So this is what the evening brings. And something hectoring, badgering me all this weekend, something belligerent, resentful. Not wanting us here. Something on the land, in the water, not settled, rising up, striking, howling. A turmoil I can't see clearly, I don't understand its desire or its need to drive us out. A short trip to the falls, the old hydro dam, tedious. Too many snowmobiles. There is poison here in this forest. Spirits everywhere. Trees ice snowbirds. Somehow it isn't right.

What is it that is off, the energy, focus, something. Can't find the words. *What did you see?* Trees ice water rocks snowmobiles' tracks water rocks. Rotting snow. Spring coming. Dog tracks, human tracks. *Not seen?* The horrors of nuclear waste. Cannot drum it, something is off, has been off since I got here.

But yet, such beauty. Yes. Restful. Too much something messing it up. Pinawa. What the forest gives. What dream is it that you withhold? Dream of the middle world – tired people, distracted. No time left.

February 6

Sleep, and not finding an answer yet to it, the absence feeling. Not knowing how to fit here. Why I am here.

What is it that has been off all this weekend, the something not aligned? Still not clear. Just me? No clear aim? But the sense from day one of defending self in spirit, though there was no attack.

What will forest tell me about water, silent in the snow, ice. Water everywhere, all white. People live on the water and in it. Contaminate it with nuclear waste. Command it, and yet are not controllers. Foul it. Waste it. Have worked with water today, a shower, coffee, food. Walking in the bush, along the trail, and we have no entry to the aquifers. Not paying attention, water pouring down in the sink.

Camp Morton, Manitoba, on Lake Winnipeg
May 31

The creator of the Water Project calls me again: do I want to go for a weekend at Camp Morton? There is room for another voice, another listener, another learner, another poet, another drummer. I say yes, yes, back to the bush, yes please. Camp Morton was once a Catholic summer camp on the beachfront, located on the southwest corner of Lake Winnipeg. Now it is a rentable resort, with new cabins, even a line of yurts along a rapidly eroding shoreline cliff. They will have to be moved, we are told. But that is in part why we are going there, to see what is happening to Lake Winnipeg.

This trip brings a community group wanting to learn about water: young families, seniors, songwriters, spirit walkers, artists. This time, we will have a special presentation from the scientist heading up the Lake Winnipeg study aboard the *Namao*, a research ship that plies the lake all sailing season, taking its measure, the measure of its health.

No Internet here, no coffee, no tea at the ready. We are assigned small cabins in the bush, each with a veranda, a firepit. Working and eating are communal. This is a vast piece of land for a summer camp, wide lawns, an old chapel, a crumbling stone wall. There were beaches but the lake has risen, trapped in its shorelines, trying to make new ones. Lake Winnipeg wants to move back out, become Lake Agassiz again.

Greeted the forest, the lake. Again meeting the land as lake and forest. We are not alone here, spirits everywhere, small birds singing, coming back to nest, the light green leaves of deciduous trees unfurling, only just. Birch, poplar. A few spruce here too, we are at the edge of the boreal, the outskirts, moving inward. Still some scrub oak. A few native chokecherry trees, nearly starting to bloom; their blossoms have a heady, exquisite scent. They are all swaying in the high wind tonight, a cold rain lashing. Patter of the leaves and the leaves sounding like wind.

Walked what should be the beaches here, but waves high on the lake, smashing up to the shoreline. Two white pelicans on the grey water, in the sun. Choppy waves, whitecaps. Stormy night coming. Found a feather, too big for a goose, yet not an eagle. But a gift. And three sticks for wands from the beach, stones smooth as young skin. And one that sparkles with silver. I will give them away. Gifts of the forest and the lake.

Long washed by the lake waters, the species of the driftwood sticks, wands, are undistinguishable. Found them along the shore, smoothed by waves, no bark, now light to the hand and showing their beautiful bones. Already so many gifts. And the day not yet done.

We are not connected here, no Internet. Something good about that, but the addiction to connecting is asking, where is everyone. They are outside, they are inside. Spirits of the forest, guardians. Everywhere. So will drum, in the cold and rain.

Calm and absolute happiness. Wind picking up, rain whipping. Lake giving at the shore. Calling the soil and the sand back and forth. So long, so wide, this lake. Holy in its giving. What is left of the sea of Lake Agassiz. We live on the lake bottom, farm on it, and make homes here together with what is left of the other animals. The old lake bottom now colonized by plants, micro-organisms. By fungi, root systems. Lichens, bacteria, algae, tiny rock eaters. Coverers of trees, mosses. Colours without equal. Light catchers. And the white, white bark of the birch.

Speaking then to the trees. No, listening. And listening to the lake. Lake says, *free me*.

What's left of Agassiz after the great lake drained and washed across the Atlantic, rolling on to create the Mediterranean Sea. Burst its boundaries and sped across the ocean. What's left of it, the lakes, Winnipeg, Manitoba. Shallow, limed. Living in the lime and granite of the old mountains, the Precambrian Shield. Oldest mountains, it is said, in the world. Maybe. What do we learn from all this. Ancient the land, but the landscape not so old. Changing.

Forest eats the rock and makes it land, soil. Calls in the water, holds and moves it, rivers, pools, lakes. Groundwater. Breathes out air as it grows. Calls up fire when it needs to renew itself. Home to the spirits that guard it, keep it alive, protect it. How to protect against rapacity? It is one thing to share, another to just take.

My cabin looks south, toward a stand of birch, poplar. There is a pink granite stone, square, maybe four feet long, maybe three high. Like an altar, and alone in the stand of trees. Even so, you have to fight a bit through brush to get to it. Wanted at least to touch it, greet it, sitting there by itself, dropped by a glacier. So an altar, perhaps to put a gift of some kind on. Perhaps it will tell me. Already the stand is familiar, some dark trees, some white, darkened by the rain. We shall see, we shall see.

We are all staying in a circle of log cabins in the forest, in the wind. Shall we ride it, walk the treetops with it, a feather in hand to lighten us? Will it make us hollow-boned like birds? What pathways to walk up there, or to dance on. What power in the tossing of the trees.

I have a good feather from a goose, a bird who knows the riding of the winds. Or a pelican. She will walk me with her, and how did the feather come to be where it was in the woods? Dropped from the sky en route, in flight. Riding the wind. Oh birds. Oh dragons. They all do it like it was nothing, like it was walking the earth for a human. Or swimming like a fish. Do the spirits get cold in the trees? Looking in the window to see what are you doing. Asking, what are you doing here? Asking, who are you? And what will you say, my name my name my forest name. Who are you, what is your forest name. I say, we will speak for the trees, sing for them.

So we do, we all do. Make music. Sing, play instruments, drum, rattle. Not enough, no, not enough. Forest is thirsty for music, lake hungers for song, story, for the long fluid lines of breath and voice as the poet speaks. It will not, it will never be enough, even the elegant graphs and schema of the scientist, not enough. This lake has too far to come, is helpless under the sun, inside the wind, against the mouths of polluted rivers that feed it, and the earth that shucks its poisons down the banks.

Voices of the small songbirds, not in the wind but in the sun. They ask, are you afraid of yourself? Spirits look in every window. Sing for us, speak, they say. You with the voice, with the drum, you with the body that sways. That dances. Why won't you sing? So I sing.

Bundled against the cold, two shawls, two T-shirts. Hands warm with the writing. Moving on the keys.

You will not wonder at this: trees speak. Everywhere, small dragons, their guardians, you can feel them, almost see them roosting in the high branches, hear their forest voices rolling in the creak of branches, trunks pushed and pulled by wind. The land is green with them, trees. If they are twisting in a breeze they are dancing. Movement everywhere. Why are you silent, listening, watching? I am writing. Feather near me. Stone and shell not far. You make an altar of a small home like this. A temple. *Look at us. Stone, feather, shell, leaf. Twigs, sticks. What will you give back? For this? What is asked for? A song. Will you make one?* There is someone resting nearby, outside.

Song for the forest, song for the lake. *More. Another.* Song for the spirits, song for the wind. Song for the water, song for the shore, song for the birds. *More more more.*

I take my drum outside and drum with the patter of the leaves, the swell and ebb of wind voice. And like the wind, the leaves, swift beats, the feet of small creatures running. Small birds in their wing flutter. Leaves batting against each other. Rain patter. It seems like someone is sitting on the granite stone, and I sing a slow song. No words, just melody. Quietly, as there are children next door sleeping, or trying to. No words, just a melody remembered, or taught. From the land, this one. And someone sitting and listening or teaching. And more tomorrow, more tomorrow. Walk the lake tomorrow, shoreline with the drum. See what songs it gives and what songs it wants to hear. Sing out. A big ship on the horizon in the haze. What song is that, what song is that?

And whose? Just wind now and a cold one. Lake rests in her bed shackled by the poisons fed to her. How can she keep and swallowing this? She will die of it. *What will you* r her and take her mourning back, repeat it, repeat, sing it back. *Now we will say again. Give back. Put healing in. Tell the story. Do what you do.*

June 1

Cold night, high winds, rain. Morning cold again, clouds over the rising sun. I walk to the lake bundled in shawls, hoodie up. The lake lashing in the wind.

Small birds in the trees, singing, darting about. A little suitor, jumping from branch to branch, fans out his tail to attract females, displaying his fine feathers. The songbirds are on their way back to the forest for nesting season, to catch the insects and fatten up for the autumn run back south.

We hear that there were high winds and rain in the city last night. Bad weather again, and we are wondering if any trees were broken in the gusts. Last night, heard a big crack in the copse beside my cabin. A tree breaking in the forest. Broken trees here, everywhere. Yet they grow again, pick up their heads and persist upward, make new branches, pull up a new trunk from a torn root. Resilience of these beings is astounding.

Wash my dishes in the lake, scrub them with sand and wait for a wave to come for water. Tricky, you don't know how far the wave will come. I end with wet shoes, but clean dishes.

Small bird sings again. Calling for mates, staking out territories. Every now and then a chase. Lake, forest, home for all. No bugs yet, but they will be here soon. Standing water everywhere from the constant rains. Little swamps in the forest. And frogs, frogs calling in the evening, the wet season gives them homes, places to put their young. Spring indeed.

Ground too wet to walk the trails, soggy with rain after rain, we stay on the high grass banks, the stone roads. Frogs sing their courting songs from the puddles, the little swamps by the roads. We talk our way back on gravel paths. Sun finally allowed out by the clouds. Wind is light, a sparkle on the lake. Haze of the other shore.

We play a game, or maybe it's a quiz, "What water do you love most?" And tease out who can most love one kind over another. What water is, does; where it lies, how; and where it rushes and falls. Where it pelts down as rain, bubbles up as a spring, lies still as an aquifer. Swamps, bogs, water stagnant and water moving. Water solid when it is ice, and water like air, a vapour. Water, brackish and sweet, cleansing, life feeding. Salt. Sulphurous.

And if I am in a warm lake swimming, that is my favourite. And if I am drinking cold well water, that is my favourite. And if I am in a bog, a marsh, a swamp, and the water is lying in pools being cleansed by plants, animals, then it is my favourite. If I am in a rainstorm, that is my favourite. If water is dashing down rocks, fall after fall, black and foamy, that is my favourite. If it is a soft little copper river, silky with tannin, that is my favourite.

Looking again at the altar stone, like a pink granite bear lying in the trees. Its mica sparkles in the sun. And two eyes. Faces, runes in the stone, sun at certain angles. Draw some, find the faces in the stone. Go to touch the stone, to sense it with hands. Narrow at one end, wider at the other. A sitting stone, a dancing stone, who seats themselves here, watches? Put out tobacco, for whoever owns the stone.

Stone carries spirit, expresses its own past, we know all this. Face of history.

A biologist comes down from the big research ship, gives us a presentation about the lake, how it is overfed, in difficulty, in peril. Wetlands gone, rivers straightened. Bad land management. Ruined land. And the farm nutrients overfeeding the water. What is to be done. Forgetfulness. Disconnection. Poison and overfeeding. No cleaning systems left. No government money to fix it, all gone to ⟨…⟩, she tells us. How do we make any sense of this? The ⟨…⟩ e me.

Keep seeing new faces in the altar stone, the bear rock. One with war paint. Another opening its mouth, small faces within larger ones. Turning light reveals them. Who dances on this stone? Who sits on this rock?

I want to take the opportunity to write, to walk the shore. Wet shoes, stones. Sand and waves rolling in from the wind. Will I want to leave here at all when this is done? Seeing the lake against the light blue sky, dark with waves pulled up by the wind. Air moves it, gives it oxygen, roils up the shallow bottom of the lake. Species and species of fish down there. Food for a lifetime. Who eats them besides us? Bears, gulls, eagles. Other fish.

And yes saw eagle chased by a gull. Eagle flapping nonchalantly, gull screaming. Eagle flew back over as if to say fuck you. Dark feathers, white head, a bald eagle. Beauty of a bird. Herons and pelicans, the glide of them, the pelicans in twos now that it is nesting time. White, white against grey waves.

You see then the faces of those lives, rock-speaking faces, speaking with faces, storytelling. Who they were or are. Pelicans gliding on wind, wings wide. We are nothing to them but competitors for the fish. And they feed well here. Pickerel especially. And these feed from others that feed from others. All beings eating each other, lending their bodies to make new ones.

What do you want to say? Only the song of little birds in trees calling for mates, frogs singing for mates. When there was a beach here, what singing! Sun through the light green leaves. Apple green, lighter, more full of light. A few white blooms on shrubs, something to make berries, bear food.

How tall and thin the poplars and birches are, reaching up beyond each other for sun. Need to get enough light to grow each year. Move into the great parliament of trees. Push out the younger ones. Everybody gets to try.

New moon, a calm night, no clouds, cool and sun still stealing from the shadows of evening. First of June, new month, and how often does this happen? A veil of wind. Crows call out a new child from the nest. Who dances on the bear stone? What is the jewel you wear? Is it the jewel of the forest? Crow's parents flutter and squawk, take the young one out for the test. Shower him with crow kisses as he walks out, pulls himself from the comfort of home, the high nest.

Looking for the spirits that dance on the bear stone, and no not happy with the hand drawing, the eye trying to see. A face, wind curling, no dance. *But you asked and this was the answer.* The hand reaches out, gives and receives. *Do you want to know what it is?* The name. Find it. Is there one? *Always. Find it now.* There is a light. A bird, a tree. *Don't look so hard, it is not far.* Not sure I need another name.

Wanting to place words down. *Why all the words?* Why not? Find a reason to do this. Bear stone. Happy drawing it, but now it's chilly, and my hands are cold. *Lake is speaking and you heard. They all did.* Science tells what the story is, and this story is wide. Layers above and below water. Layers of life. Of feeding. And what do we do with this.

Ready to go home now. Returning tomorrow, and it was like this last time. *You are no prize but not bad at times.* What then have I found here? The lake especially, Karen Scott and her ship sounding the water three times a year. Discovering its status. Takes a ship to do this body of water, and the bottleneck of the narrows. North and south lake.

What can be done? This is an old lake bottom. We live on a flood plain. We live partly in water all the time.

Still light on leaves, sun in the west going down. Light and sun, and oh give us more tomorrow. Wanting to sleep, rest for the eyes and mind. But heart too angry for the forests and the lake. How did we come to this, to this, where we neither know nor care where our food comes from. Stewarding the land does not work. No understanding, not caring to know. Hard to leave this place. The bear stone, the crows, the heaving lake calling us to save it. Free it from the poisons we put into it.

QUESTIONS TOWARD AN ETHICS OF VIOLENCE

When stars burst, new galaxies become possible. When the planet shrugs, shifts, as she always does, new land is formed by the fire of volcanoes. When waters rise up, great floods, tsunamis cleanse land and replenish soil. When earth moves tectonic plates under oceans, quaking and shuddering, mountain ranges crumple into the heavens. Old forests and prairies die, of necessity, by fire, to make room for new growth, enriched soil. Tornadoes, hurricanes bash the land, wash out debris; windstorms move soil, creatures, seeds, from one place to another.

And life, oh life, yes, life feeds always on other life: it is a chain of killing and eating, from one end of the biological spectrum to the other. Genetic material is selected, chosen, and passed on by means of intense battles among males of many animal species, sometimes to the death. One could go on.

All this is utterly violent: but without violence, of course, there is no universe, and there is no life. Violence is the power of creativity, continuance, survival; the power of rectification and rebirth when a forest goes stagnant, or when the shores of a river are depleted of nutrients. It is the power of survival for all forms of life, for whether they are eaters of plants, or of other animals, or both, every form takes its life from the living. And it is the power of balance, too; it is violence that drives, even ensures, the flourishing of diversity in life forms, as species keep each other in check, and in good health.

Yet this natural violence does not in our time have an ethical status that I know of. In the West, we no longer have gods to carry this for us, to demonstrate it, to do what is necessary for creation and flourishing. And we no longer, as far as I know, possess a comprehension of natural violence that recognizes it as inherently, profoundly ethical.

So I ask myself this: do we have, anywhere, an account of violence which would address our understanding of natural violence in ethical terms?

Here's how I came upon the problem, how the question arose. For some years I was studying eco-feminisms in relation to Shaktism, which, for those who are not familiar with it, is the Hindu tradition where the Goddess, or Devi, is supreme deity, and is identified as *Shakti,* power. She is also identified as *Prakrti,* matter, and as *Maya,* which can be interpreted as illusion, or difference, or creativity. This theology is immanentist, and so places the divine within and throughout the world. At a certain point in my research, I wanted to consider whether Shaktism could support an eco-feminist ethics of mutuality, particularly as expressed by Val Plumwood in her *Feminism and the Mastery of Nature.*

Mutuality can be described as a reciprocal relationship, as one of mutual dependence. An excellent example of natural mutuality is illustrated by the return of wolves to Yellowstone Park: without the wolves, elk overpopulated and denuded river- and stream banks of flora, leaving them without trees and other vegetation. This in turn left no place for birds and other small animals to find homes and food. Without shade and cover, and the detritus of trees, aquatic animals could not live in the rivers. Once the wolves were reintroduced, and elk populations were thinned, flora could again grow on the riverbanks, in turn providing food, homes for small fauna, and encouraging, bringing back the flourishing of a diversity of life forms.

From an eco-feminist point of view, and certainly from Plumwood's point of view, that mutuality would extend to relations and reciprocity between humans and the rest of nature. Yet already there is distance between them, at least in our current cultural narratives about nature, and about the place (or not) of humans within it.

I ran into serious problems with my study of eco-feminisms. Because mutuality is presented as a way for humans to relate to the rest of nature, it does not in general give a mutualistic account of nature itself, including the essential role of violence in natural mutuality. For example, in Plumwood's assessment, mutuality is presented as a psychology and she repeatedly uses such terms as kinship, intersubjectivity, and relationality, but in service of how humans relate to what she refers to as "earth others." Nothing wrong with this, in and of itself. It's an excellent model, and relationality and mutuality and kinship work very nicely with an immanentist theology such as Shaktism, where, since all things are divine, all things are related and participate in the matter, action and subjectivity of the Goddess. All well and good.

But in the physical universe, there is no survival, no flourishing, without a sharing of space, form, time, energy and matter. And that sharing includes, even depends on, violence: the sharing of energy and matter, for example, requires the killing and eating of one life form by another. There is no room here for "feel good" psychology, there is only lived life. A mutuality in the lived world of nature would recognize that we humans are food, too. It would include an understanding of the violence we do to nature ourselves, to other beings, just by taking up space, by competing for resources, for land. Mutuality of humans and other natural beings includes violence, by necessity, and I wonder in what ways do we recognize this? Did the people who rid Yellowstone Park of wolves in the first place understand the violence they were doing to an entire ecosystem, to countless species of plants and animals, in order to make it "safer" for themselves, for livestock?

There is no question that violence is a fundamental part of the nature community, and its ethical role is positive, because it is essential to life. It is the force, the power, the *shakti* of creation. We know this from science, but do we know how to understand it ethically? We call it a necessary evil. We anthropomorphize nature as wrathful or unforgiving. Or we call its violence unavoidable. I have even heard people suggest that predatory animals just don't know any better, or that animals and plants don't have or need or live in a world of ethics, or, they have no choice (thus reducing them to automatons, machines). This lacuna of ethical understanding is created, in part, from us not recognizing ourselves as animals, from setting ourselves apart ethically from the rest of the natural world.

One way to escape addressing the question is by identifying the quotidian violence of creatures fighting and eating each other to survive, as ethically neutral. Even, as evidence of a "fallen" world. Such explanations derive directly in Western culture from our inherited theology of transcendence and hierarchy, where humans answer to a "higher" ethical order than other beings, are inscribed as "special" to the divine in our creation narratives. As a consequence, ridiculous moral ideals such as the lion lying down with the lamb are substituted for the natural order: it is the lion's proper place to eat the lamb, not lie down with it. Violence is not separate from the creative processes of the universe, nor from its maintenance. Acts of violence, large and small, occur at every moment, in every corner of the known and unknown universe. How do we understand this ethically from within our inherited narratives?

First problem

In contrast to Shaktism, ecological feminisms, as well as other feminisms, have given the term violence an exclusively negative signification. I actually looked up the term in all the books I read by leading eco-feminist writers, and found it used only one way: it is applied to human behaviour, and it is a sin. Period. The term has been masculinized by feminist thought, and rendered patriarchal. I will not argue that human uses of violence have not been wrong, harmful and shameful, horrendous, and frequently patriarchal. However, the term violence in feminisms, including eco-feminisms, is deployed exclusively as descriptive of a tool human males use against a collective of "others." This structuring of the term fails to find an ethical location for violence that is in, and of, nature. I simply noticed this failure when I considered eco-feminisms alongside Shaktism and science.

Like many other writers, Plumwood has called for an eco-feminist ethics based on mutuality, identifying mutuality as a site for the relational self, and she has called for an ethics of care. She argues that an ethics of care based on mutuality (humans living in mutuality with the rest of the natural world) will generate the flourishing of all (humans and other creatures) as its product. So far so good. In the example of the Yellowstone wolves, it was human intervention (ridding the park of wolves in the first place) that created the problems for the rest of the ecosystem. And it was a human intervention of care by returning wolves to the park, and "living with" them, that restored the balance, allowing the whole park to flourish.

My finding has also been that eco-feminist understandings of mutuality as an ethics do not address the mutuality that biology calls mutualism, the mutualities of predation and parasitism, as either ethical or non-ethical. Nor do eco-feminist writers, at least those I have read, consider the ethical necessity of predation and other types of violence to the enterprise of survival and flourishing. Yet for both participants in a predatory relationship (predator and prey), this is a "mutually beneficial association between different kinds of organisms."[1] It should be noted that predation, along with colonization, domination and war, all exist to varying degrees in non-human nature. Biological accounts of natural violence are often missing from eco-feminist discussions of nature, alas. It is as though the theorists do not want to acknowledge that nature itself is fundamentally violent. But it is precisely this violence that constitutes the dynamic for natural creativity and sustenance, and the necessary transformation of matter. Is this not ethical?

Even though some thinkers (Eliot Deutsch, James Lovelock, Roger Lewin, for example) have argued for a nature with intentionality, as a superorganism and so on, in general, Western thought lacks a description of nature which would make it a moral agent, and tends to assume that humans are the only embodied moral agents in the universe. This in turn generates moral hierarchies within nature itself (assuming humans are part of it), which have spilled into eco-feminist thought. It is as though there is a fear of opening the question of violence itself, and acknowledging that it is, at least, morally ambiguous. The result seems to be that while certain kinds of violence are deplored in humans, violence among other creatures is ignored ethically.

The term "violence" may be understood and used two ways: from its word source "to violate" it is used negatively, as unethical and harmful human behaviour (violence against women, against nature, against races and so on). It is also used simply descriptively: storms are violent, the big bang is violent, volcanoes are violent, imploding stars are violent and so on. My reading suggests that these last forms of violence are considered ethically void, they are just nature in the state of creative disequilibrium. These activities are not actually violating anything, as they are considered mindless; even so nature is often referred to negatively as an angry and destructive mother when natural violence affects humans.

Second problem

The second problem is the question of victimology. Let's go back to the example of the wolves in Yellowstone Park. Typically, where there is predation, too often prey animals, in this case the elk, are characterized as their "victims." Likewise, and more pointedly, the ranchers whose stock may be "game" for wolf predation might see themselves as the (economic) victims of those wolves. And where there is a victim, there is always, must always be, an evil perpetrator. I mean by this that once the term victim is used then who- or whatever is at the victimizing end of the stick becomes a target for judgment, human justice and (far too) often extermination. Yet the by-products of the predator-prey relationship between wolf and elk (a mutuality in itself) extend past the mutuality of wolf predation on elk, and on to the flourishing of riverbanks and rivers, communities of plants, birds and water creatures, which in turn feeds another multitude of creatures, and the soil itself. Victimology can be a short, narrow and self-interested lens. It can also be tainted by a false, or at the very least, shallow, empathy (for the elk, for example).

I would like to define true victimization as the product of the intention to do harm without justification, as, for example, in a rape or a murder. While it is clear that the wolves intend to bring down the elk for their dinner, it is not clear that they pick off calves or lambs with the intention of doing the ranchers economic harm. Perhaps a good way to define genuine victimization is as there being no mutuality between a true victimizer and a true victim.

Because there is no holistic description of natural violence in their ethics of mutuality, eco-feminist accounts of violence are also oppositional, and violence is ethicized as bad and wrong. It also paints violence itself as male, human and non-natural, creating a kind of ethical reversal, and reinforcing the alliance among women, nature and other oppressed beings, and, by consequence, generating an identity of victimhood for those categories. I don't want to go into all the implications of the ethics of victimology here, and likewise I do not want to minimize the suffering of those who find themselves at the wrong end of the patriarchal stick. What I am suggesting is that to start ethics of any kind, one cannot always start them from a position of marginality or exclusion, or of victimhood, in particular, a position of claiming victimhood where it is not applicable.

The dilemma of victimhood fails to construct an ethical location for violence in nature. We still describe a deer as the victim of wolves, for example. Or a person as the victim of a disease. If one is part of nature then one is at all times doing violence to something; something is always going to be our "victim," be it a mosquito or a bacteria or a carrot or a nice bit of lamb. Where do we place ourselves in nature such that it is not unethical for us to live? By not accounting ethically for violence in nature, eco-feminisms are reinforcing the culture-nature split it seeks to dissolve because it sets up separate ethical accounts regarding violence for humans and for the rest of nature. Furthermore, it does not always, or altogether, account for human violence in the quest for survival (food production, for example). Some eco-feminisms address this question by recommending vegetarianism, but do not acknowledge that this is also violence, in this instance, against plants instead of animals. This sets up another hierarchy of being, where animals are higher than plants. A series of ethical backfirings occurs therefore as a consequence, perpetuating the types of dualisms and hierarchies that eco-feminisms wish to erase. Survival is not separate from other ethics.

One could suggest that this problem exists in part because Western theology has traditionally excluded violence from God. Except for just wars, punishing bad humans and, of course, Jesus, who was himself a victim of human violence. Externalizing violence from God has created a problem of evil, as well as the dualism of nature/culture: in this construction, God is not nature, God is not tainted by its processes. The problem of evil sets up these oppositional ethical approaches to which eco-feminisms are heir. It is therefore oppositional itself, and cries out, on behalf of all the oppressed and excluded categories, from the margins of the dominant discourses. I am not saying this is wrong, I am saying it is oppositional. Eco-feminisms correctly identify both the wrongness of these dominations, and the connections among the categories of the oppressed, but they do not perhaps recognize that they may be working from Biblical assumptions themselves, though in opposition to them, and may be generating the same kinds of dualisms as a result. Certainly, they work from a dualistic position of separate ethics for humans and the rest of the natural world, even from within an inclusive ethics such as mutuality.

Third problem

The third problem is dualism. Or rather, a whole set of dualisms. Dualism between humans and the rest of nature, dualism between spirit and matter, between the sacred and the profane. Certainly in the West, these dualisms are firmly rooted in Biblical culture, in Greek philosophy, Cartesian thought and continue right through to the present. We work from assumptions about a "specialness" of humans over other animal life, we privilege spirit and mind over matter, the rational over the intuitive, text over orality, and we especially privilege human needs and wants over all else. We are able to justify this in part because we have desacralized the natural world, removed sacredness to the realm of a remote deity. And of course, where applicable, to human souls.

We work, then, willy-nilly, from narratives rolling out from theologies, and even attempts at inclusiveness ethics, such as eco-feminisms, still fail to elude dualism (here, separation of humans and the rest of nature) and to account ethically for natural violence. Eco-feminisms do not err in identifying the all-too-frequent deployment of violence by humans (wars, slavery, oppression, ecological rape) as evil. But, it is limiting the availability of the term violence to a narrow ethical application. Doing so reinforces the dualism between humans and non-human nature, and between culture and nature, as it sets a different standard for natural violence than for human violence. We do not seem yet to have a narrative that can address natural violence as ethical, not in the West. This is why I looked to Hindu Shaktism for answers.

Even though eco-feminisms and Shaktism can be viewed as social critiques, Shaktism is much larger than eco-feminisms and it has more epistemological tools. As a religion, it has at its disposal a larger array of lenses for looking at the universe. Shaktism can offer an account of nature which bridges theology and science: the divine generates the physical world out of itself. It can, therefore, offer an account of natural violence within, and not outside of, a moral or ethical universe. Violence can be accounted for ethically without making either nature or humans "other."

In Shakta literature and iconography, the Goddess is often depicted as dark, violent, bloody and unmanageable. In the *Devi Mahatmya*, a central text of Shakta Tantra, Devi and her *shakti*s roll over battlefields with astonishing ferocity. Kali arises as Devi's anger, she drinks blood. In the iconography of the *Ten Mahavidyas*, or Great Wisdoms, goddesses appear in transgressive forms: old, single, emaciated, fanged, self-decapitated, bloodthirsty, ugly, sword-wielding, wearing skulls or decapitated heads around their necks, stomping on male figures and so on. What ethical role does their/her violence play in the unfolding of Devi's universe? David Kinsley[2] has argued that such goddesses reflect nature itself – it is red in tooth and claw, and there's no getting around that. Is Devi's violence a kindness? Devi has no male consort; she is a free moral agent in the wielding of power. When Devi goes to war, her battles are about cosmic crises, re-establishing a moral order.

Shaktism manages to acknowledge the violence of the natural processes of creation, maintenance and destruction while at the same time ethicizing deliberate uses of violence to restore cosmic order, *dharma*, the ethical. Because Shakti herself *is* nature, there is no separation between the sacred and the profane. Since the phenomenal world is her body, *prakrti,* then it is, all of it, by definition, always, and already, sacred. In this model, there is no such thing as the profane, and, therefore, no dualism generating these two realms. Do we need to (re-)extend our understanding of sacred space to include the whole of nature in order to have ethical consistency?

The Shakta cosmos is the divine self-embodied. Creation begins with a deliberately generated disequilibrium within Devi herself. Violently. Equilibrium is broken, disrupted. Devi chooses to do this. She is a moral agent. How is violence to be understood when it is part of Devi's self-revelation? And if it can be said that the "dark" aspects of Devi are both descriptive of nature and have an ethical meaning, and if the bottom line in nature and for humans is always survival, can violence therefore be thought of as good? Compassionate?

Shakta narratives of violence (including visual narratives) are about salvation. How violence is used in the exercise of power is at issue, and who is using it and why. The Goddess exemplifies violence in iconography as Kali, Chinnamasta, Dhumavati – revealing the violence of time, of the cycles of life and death. In this sense, violence both reflects nature and also saves the world from imbalance or bad use of power. The ethical knife-edge for humans is the same as it is for Devi – not violence in and of itself, but how and why violence is exercised, the intentionality of violence. For the Shakta, the universe is intentional, it is Devi herself.

We have nothing like this in Western culture, not anymore. We have long since lost or forgotten our ancestral narratives of divine partnership in the making and working of the world; at least, those of us whose ancestors are European. The gods no longer walk the earth, no longer give their torn bodies to make the heavens and the earth, infusing the cosmos with their own blood, their divine essence. We have nothing like Devi in our current lore, our texts, sacred or secular. We are, in a sense, bereft, and on a long climb back to that old understanding. I have noted above the work of eco-feminists[3] and scientists like Lovelock and Deutsch, who have struggled toward an account of the world that reinstates intentionality, even the sacred, to matter. Yet none of these offers Shaktism's bracing (and embracing), holistic account of nature, an account that acknowledges, even celebrates, the essential, creative, ethical role of violence in the world, in the cosmos.

So then, let's return, to the wolves of Yellowstone Park. Since their reintroduction to the park in 1995, their numbers have grown, but not out of proportion to the needs of the ecosystem. They live on the edge, their numbers and their health always regulated by the availability of their prey. And the prey – primarily elk, now themselves regulated by the presence of wolves – no longer decimate the landscape by virtue of their sheer (unnatural) numbers. For the wolves, there is often hunger, illness keeping their own populations in balance. There are fierce battles for territory, for food, for the right to mate, for leadership, sometimes to the death.

Whether or not wolves intend to make Yellowstone's riverbanks flourish with abundant life by keeping the elk in check, nevertheless, that is the outcome of their natural and rightful presence, their predatory role. It is their intention to survive that drives this engine, and that, survival, is ethical for all beings. And it is their willingness to take violent action on their own behalf that honours and manages the mutuality of relationships with all other beings in their environment, plant and animal. The presence of wolves regulates, too, the populations of coyotes, and has driven cougars back to their natural mountain habitat. As if that were not enough, wolf kills feed many other creatures, like eagles, ravens, bears, magpies, wolverines. The tumble-out of benefits toward the flourishing myriad of other species reaches all the way to the diversity of plants, and to the health and vigour of waterways. Not bad for one predatory species. Not bad for the Big Bad Wolf.

Notes

1. *Merriam-Webster*, s.v. "mutualism," accessed April 30, 2013, http://www.merriam-webster.com/dictionary/mutualism.
2. David Kinsley, *Hindu Goddesses: Visions of the Divine Feminine in the Hindu Religious Tradition* (Oakland, CA: The Regents of the University of California, 1986). Published by arrangement with University of California Press.
3. In particular, spiritual eco-feminists, such as Starhawk, Charlene Spretnak, Carol J. Adams, for example, have looked to the past for an appropriate divine model.

Second Spring

FOREST DIARY, FALCON LAKE, MANITOBA

April 14

Seems when I feel a call from the forest, someone answers, offers a workshop, a bit of teaching, a few days by the lake to think of the nature of water, something. This time I am asked to be a guest teacher for a weekend writing workshop at Falcon Lake, deep in Whiteshell Park, the Precambrian Shield, its boreal forest, the lakes, lakes, lakes.

We drive in to the Whiteshell late in the afternoon, in the rain, and pull up to the resort where the workshop will be held. Sky is pearl grey, light diffuse from thin mist. Patter and rattle of rain on the roof of our chalet. We are given rooms that have the names of forest wildflowers. Mine is Lady's Slipper, the little yellow pocket that insects love to dive into. It is mid-April, and the deciduous trees are beginning to fuzz up. Conifers are already lightening their needles from the dark greens of winter, extending their brilliant spring-green hands. Catkins push out from some of the bare birches. Birds are busy in the trees. An incongruous forsythia blooms cheerfully yellow by the chalet door.

We don't see them, but we know bears are foraging, mothers with cubs. Too close here to humans. Along the lake edge the forest rises, drops sharp, trees holding to the rock edges as if it were nothing for them to cling by their claws. Dense little forests of new shoots on the forest floor, still littered with last year's leaves – those not yet digested by the land, the microbes. Pale blue-green lichens cling to trees, luminous greens of mosses lush with all this spring rain.

To the north of us, just north, and down the rocks is the lake of the falcons, and I can't take my eyes off it. The horizon of forest makes a dark blue scribbling on the long curved shore, the grey-silver water, deep green islands. Enormous sky. The owners of this resort have built a deck that looks out to the lake, so you can lie back into its beauty and never stop looking. I take my drum outside and sing with the land. How can you not.

The air here is astonishingly clean, deep with plant scent and oxygen. It's an air forgotten in the city, even when it rains. I say hello to the forest, trees wave back in the wind, bouncing their branches like hands. Greetings, greetings from forest to city, city to forest.

Women have gathered here for a writing weekend, workshop after workshop in the crafting of it, and the joy of it. Most of the women are from nearby, from the forest, but some have come from far to the west; wanting to get away from families, from work, and into the healing of the boreal.

First night, after dinner, we are entertained by three young musicians, players of banjos, guitars, a fiddle. I ask to drum along, and they are kind, they say yes. Soon enough another drum comes out, spoons, the thump of feet. We are the little orchestra of the bush. A girl gets up and dances jigs. Luckily there is plenty of wine.

When the music ends it is late evening, cool, about ten degrees Celsius. Four of us pull on our bathing suits and slide into the hot tub, let the warm water soothe our limbs. We make mist with our bodies, and a little drizzle cools our faces, our shoulders. We are looking up into the deep night, still clouded, up the white stems of birches hanging catkins at us, and we are breathing the cool air into our souls. How could we know we had longed for this. Water, air, scent of the trees in spring. Silence we forget in the city. Quiet. Everyone sleeping but the land.

April 15

I wake up at dawn in the Lady's Slipper room, and it is not quiet. Birds whose names I don't know are calling in the trees, from the water. Ducks quacking themselves silly on the docks. Crows, crows, and one call from the raven, keeper of the forest, the land. Mist on the water, the lake waking itself from the night's stillness, little islands looking at themselves in its surface. We don't know what they think of us, our important lives are nothing to them. Their business is life itself, creating it, breaking it down, land and water, the deep mosses, fish surfacing for light and the coming insects. Not here yet, the fish and insects, it's too soon. Forest is still waking up from winter.

Forest reminds you it will not bend for you. Do we wish to hear its narratives, do we wish to be its students, its partners? Difficult to find our way back to that, the partnership, the respect, the lore of living in it, with it, from it. For us of European descent, the lore is too old for us to reach anymore. We lost it long ago. Had nothing of it left to bring with us to the new land.

We have to start again, start with the ears, the eyes, the body. Manners. How to ask a tree if it will consent to be cut down for a house, for a boat. This is the old law, natural law. How, when we are so distant from it, do we find our way back? Straightening rivers in their beds, we do not ask them first, nor the thousands of beings who depend on the rivers to meander. We do not offer a gift for the shifting of land and waterways, the building of dams, the pushing out of the biota from its oikos. We used to know this, we used to know the law. Crow still calls it out to us, in the cities, in the forests. We don't know how to hear it anymore, and here and now I am deeply aware of this deafness.

The story goes that dragon's blood gives you the power to hear the speech of birds, birds who are the messengers of the gods. Dragon is the land, to some, but we long ago lost the lore of the gift of dragon's blood. So I hear the birds in the morning, trilling, quacking, croaking songs like rusty nails, and I don't know what they are saying. What the voice of the gods is. Or rather, what their message is. Their conversation, talk.

What I see is this: the land gives and gives and gives. We give nothing back. The old law is broken, flouted, forgotten. How long will the forest do this? How long will its creatures, the forest spirits, keep giving? This is the price of urbanization, and that is older than the modern world, older than the Christian world. We inherit from the Christian mythos, from a Biblical view, that humans are at the pinnacle of nature, and a belief in human entitlement to all the goods and to the giving, and then, who guards and speaks to the land, the forests? Who cares for them, talks to the spirits, who remembers how to give thanks for the gifts with gifts?

There it is, and it seems there is no end to it: the boreal forest. You look out at it from a window, or you look at a land map, you see it ringing the whole of the northern globe. You think it is so vast, so powerful, what can touch it? You can remind yourself of the forests of the Amazon, the felling, the landscape shifting into wasteland. This can't happen here. This cannot happen.

Morning here is all birds, mist, grey light. Duck at the shoreline again calling hard for someone, and another wings in from the south, morning row call. Chickadees, flickers and falcons singing, chittering from the trees. Crows again, and again one short croak from the raven. All birds, and water. Forest is insanely green, the conifers. Birches blooming in the cold and rain. Islands watching themselves in the still water of the morning, magic at that liminal moment, that liminal place where land meets water, another world open in the brief mirror of the lake.

Birch peel off their last year skins, pink new bark, the old skin almost grey from rain. I walk up the hillock, a piece of the Shield, broken by time and glaciers. Stones are starred with lichens eating them slowly into soil. How long and measured it is, this soil making. How tenuous the forest, grasping the old rocks, holding on, some pushed over by wind or their own weight. I walk in with one of the women, a hilly walk, snap of old twigs and soft moss underfoot. We are smiling all the way, talking, running up and down the tiny hills of stone and moss.

Water, always water in the boreal: rain, lake, the water held and called by trees. Water as tiny pearls hanging from the unleafed trees, strings of them. Water in the needles of conifers, water in puddles on the road. Water caught in the cups of lichens, tiny drops. Water caught in my hair, curling it.

We walk the forest carefully in the spring, watching for ticks, bears, for tiny budding plants underfoot. We are home. A bird voice calls at me, or to a mate. And I think how dragon's blood, just a little, would give me its speech, or an understanding of its words.

The women gathered here for the writing workshop have lunch together, tell each other about their lives. What women do in the forest, when they are not gathering sticks, berries, leaves, roots. When they are not carrying their water pails down to the lakeside to dip and fill, when they are not walking the hill to the sauna, wrapped against a chill, singing.

These are all women of the land. They know it, do not fear it. Some of them have lived here all their lives. The forest is not a resort, nor a holiday landscape, but the landscape of their minds, and of their bodies. They are nourished by its berries, its air. They swim its lakes. They know every bird, and the safety of trails, the passing of seasons. They are keepers of land, and they are bodies of the forest as the wolf is, as the squirrel, the bear, the deer is. I watch them navigate it and rest in it. Move about it like its extensions, like any other animal.

It begs the question then of how humans have always migrated, have walked, boated, sailed, ridden from one piece of land to another, bringing with them their tools and gods, their need for food, for land. How these humans might move from plain to forest, or forest to desert, adapting and adapting, making themselves part of the land they take, settle on, fight for. Negotiating with the people, the earth, the spirits, the creatures whose home is in the sea, the lake, the shore, the forest these humans now inhabit. How they dealt with the animals who would now be their food. Did we as invaders, settlers, coming to this land from Europe, carrying with us our Biblical narratives of human entitlement, triumphalism, did we make these negotiations? Or have we strayed so far from our own ancient knowledge, the knowledge of our own ancestors, that we no longer have the tools to make these accommodations, these negotiations, with the land?

You would not believe how fierce is the desire of trees to grow on these ancient rocks, the Shield, worn down old mountains. How essential it is for the birch to be white, call light down to the dark forest floor. You would not believe the persistence of water pulling life to itself, paying back life. Who is to say what is strength, what is not? Lichens grab, hold, eat granite into soil. They are not the only ones. But how small they are, and such tenuous partnerships of separate beings. The way the earth makes, remakes itself, in tiny mouths, the long arc of their eating.

Somewhere in the middle of the day, I am called, pulled to a clearing not far from our chalet. Someone is there, summoning. I close my eyes and there are wolves all around me. Red wolves, with their long muzzles and short fur, not the timber wolf, as you might expect. And a being, human-shaped, indistinct. *What is your name?* and I give it, the forest name. My hands are asked to move, shape, this way or that, hold, give, bless, send in energy. I feel something that reads as headphones at my temples. The question, then, *What do you give?* I say, *Words, that is my gift*. The question, *What do you want?* I say, *Freedom*. Hear: birds, drip of water, wind in trees, birds, birds. Far away, deep in. I try then to return to the middle world, *Not yet, not done*. It seems like hours, then, at last, *You can go*. So I say thank you, I walk back the millions of miles from that other world, or fly. You don't know where the forest will call you, what place inside itself, to speak and listen, nor how long it will want to talk.

My turn now to give a writing workshop, and so I tell the women about the Boreality project, how we went out, listened to the forest in four seasons, wrote what we heard to make the *Cantus Borealis*. I send the women out to listen, and then to write. I ask, how many write poems, and only one says yes she does. The rest, no. They scatter into the nearby forest, down to the shore then, dubious, I suspect, for half an hour or so. One woman comes back early, radiant, smiling, asks me if I am a hugger? She wants to hug me, she has heard the voice of a dead loved one. The forest has healed her. It does that. I tell her so.

One by one, the women return with their writing, their encounters. All but one has written a poem, each has opened her ears and listened to the forest. One woman has heard the complaint of it, about the noise of people and their machines. Another has listened to the voice of stone, telling her how its life is slow and long. Yet another encounters Squirrel who tells her she will be told the lore of the forest, its stories, when she is ready, and this girl is the one who needs to know she is worthy, a person of value. Each woman returning has listened, and has written what she has never written before. I am hearing this, or seeing it, as they read what they have heard, written down: voices of the forest come in poems. The forest speaks in poetry.

And I am thinking, the wolves led the women, then. Pathfinders, those wolves and teachers. So I am grateful to them, to the forest, for the teaching and speaking, and I am proud of the women for opening their minds and ears, for listening. For writing their first poems, most of them.

There is a big pink granite stone lounging casually out front of the chalet, and I am reminded of the altar stone at Camp Morton. So I look to see who is in it, in this stone, and face after face reveals itself, spirits or the imprint of their features. And like they were having a joke on me, one after another, some smile with mischief. You don't know who lives in a forest, who makes it their home.

Last night here, and we decide to have an old-fashioned singalong. Out come three guitars, a drum, a harmonica, and voices, voices, red wine. We sing for hours: blues, folk songs, old rock 'n' roll, turning page after page of an immense songbook. Happy, happy. And the rain still coming down outside in the dark, the lake eaten by night, birds silent and sheltering.

I think of old songs I learned, taught myself, when I was young. The forest calls up memory, my walking with it again and again, my youth. There is, there always was, exuberance, *frisson*. Everything ecstatic, loving itself. Dust, sand, forest, prairie. City streets, bars, music. And then you walk into the forest, familiar with its spongy soft footpaths, a cracking or two of twigs fallen. A quieting of the air. Mosses and tiny plants that break down stone. What have I forgotten about this? Nothing.

April 16

Raining again, wind's up, lake all grey pocks and ripples. We are taking leave of it, and the great forest that gathers it in. One last workshop, a farewell lunch and I am heading back into the city. I wait for the bus by the side of the highway, have to flag it down on its route from Kenora, Lake of the Woods, to Winnipeg. Greyhound stops with a *psssht* of its huge brakes, I climb in, settle down for the short drive, ninety minutes back into the city. We ride the highway non-stop, and the forest thins and thins, then disappears into flat wide prairie as we roll toward town.

ACKNOWLEDGEMENTS

An earlier version of "Does Nature Have Rights? Ethical Implications in Ecology" was published as "Does Nature Have Rights?" in *The Journal of Faith and Science Exchange*, Boston Theological Institute, Boston, MA, 2001.

A shorter version of "Questions Toward an Ethics of Violence" was presented at the CSSR (Canadian Society for the Study of Religion) Congress at the University of Manitoba in June 2004.

Some of the seasonal forest diaries appeared in earlier forms on *Prairie Fire* magazine's Boreality Project website as *Kate's Boreal Blog* (http://www.prairiefire.ca/Boreality/kate.htm). An abridged version of the forest diaries also appeared as "Notes from the Boreal Forest" in the Boreality issue of *Prairie Fire,* Vol. 33, No. 1, Spring 2012.

SPECIAL THANKS

To my brother, Tony Szumigalski, and sister-in-law, Carla Zelmer, biologists, for teaching us the science of the boreal forest.

To Bob Haverluck and Donna Besel, for taking me back to the boreal, again and again.

And to Noelle Allen, for suggesting this book.

Katherine Bitney is the author of three critically acclaimed books of poetry: *While You Were Out, Heart and Stone* and *Singing Bone*. Her fourth collection, *Firewalk,* was published by Turnstone Press in fall 2012. She has worked as editor, mentor, creative writing instructor, arts juror and literary creative director for over thirty years in Manitoba. Most recently she developed the text for *Cantus Borealis,* a choral piece on the Boreal with composer Sid Robinovitch (premiered April 2011). Katherine Bitney holds a Master's degree in Religion.